I0055055

Principles of Thermodynamics

Principles of Thermodynamics

Henry Medina

Principles of Thermodynamics
Henry Medina
ISBN: 978-1-64172-659-7 (Hardback)

Published by Larsen and Keller Education,
5 Penn Plaza,
19th Floor,
New York, NY 10001, USA

Cataloging-in-Publication Data

Principles of thermodynamics / Henry Medina.
p. cm.
Includes bibliographical references and index.
ISBN 978-1-64172-659-7
1. Thermodynamics. 2. Heat. 3. Mechanics. 4. Dynamics. I. Medina, Henry.
TJ265 .P75 2022
621.402 1--dc23

For more information regarding Larsen and Keller Education and its products, please visit the publisher's website www.larsen-keller.com

Table of Contents

Preface

The branch of physics which deals with temperature and heat, and their relation to energy, work, radiation, and properties of matter is known as thermodynamics. The four laws of thermodynamics form the basis for the behaviors of these quantities. These laws are the zeroth, first, second and third law of thermodynamics. The zeroth law defines thermal equilibrium. The first law of thermodynamics is related to the principle of the conservation of energy. The second law deals with the flow of heat and how it cannot flow from a colder object to a hotter object. The third law defines how all processes would stop if the temperature of a system becomes zero. There are various branches of thermodynamics such as classical thermodynamics, equilibrium thermodynamics, chemical thermodynamics and statistical thermodynamics. The topics included in this book on thermodynamics are of utmost significance and bound to provide incredible insights to readers. The book aims to shed light on some of the unexplored aspects of thermodynamics. It will serve as a valuable source of reference for graduate and postgraduate students.

A foreword of all chapters of the book is provided below:

Chapter 1 - Thermodynamics is the discipline of physics that studies the relationship of heat and temperature with energy, radiation, work and the properties of matter. Thermodynamic system, thermodynamic state, thermodynamic equilibrium, etc. are some of the concepts of thermodynamics. This is an introductory chapter which will briefly introduce about these concepts of thermodynamics; **Chapter 2 -** The laws of thermodynamics define entropy, temperature and energy which characterize thermodynamic systems and their equilibrium. It includes zeroth law, first law, second law and third law of thermodynamics. The topics elaborated in this chapter will help in gaining a better perspective about these laws of thermodynamics; **Chapter 3 -** There are different types of thermodynamic processes that affect pressure, volume, temperature and heat transfer. These processes are isothermal process, adiabatic process, isochoric process, isobaric process and reversible process. This chapter closely examines these various thermodynamic processes to provide an extensive understanding of the subject; **Chapter 4 -** Entropy is the thermodynamic quantity which represents the degree of randomness in a thermo-dynamic system. Clausius Theorem, Clausius Inequality, Temperature-Entropy Diagrams, phase transition, etc. are some of the concepts that fall under its domain. This chapter discusses in detail all these concepts related to entropy; **Chapter 5 -** Thermodynamic potentials deals with the representation of the thermodynamic state of a system. It includes internal energy, enthalpy, Gibbs Free Energy and the Helmholtz Free Energy. This chapter has been carefully written to provide an easy understanding of these thermodynamic potentials; **Chapter 6 -** Mayer's relation, Maxwell's relations, TdS Equations, Bridgman's thermodynamic equations, Clausius Clapeyron equation, Gibbs–Duhem equation, Joule–Thomson effect, etc. are some of the aspects that come within thermodynamic equations. All these aspects of thermodynamic equations have been carefully analyzed in this chapter.

At the end, I would like to thank all the people associated with this book devoting their precious time and providing their valuable contributions to this book. I would also like to express my gratitude to my fellow colleagues who encouraged me throughout the process.

Henry Medina

1

Thermodynamics: An Introduction

Thermodynamics is the discipline of physics that studies the relationship of heat and temperature with energy, radiation, work and the properties of matter. Thermodynamic system, thermodynamic state, thermodynamic equilibrium, etc. are some of the concepts of thermodynamics. This is an introductory chapter which will briefly introduce about these concepts of thermodynamics.

Thermodynamics is the science of the relationship between heat, work, temperature, and energy. In broad terms, thermodynamics deals with the transfer of energy from one place to another and from one form to another. The key concept is that heat is a form of energy corresponding to a definite amount of mechanical work.

Heat was not formally recognized as a form of energy until about 1798, when Count Rumford (Sir Benjamin Thompson), a British military engineer, noticed that limitless amounts of heat could be generated in the boring of cannon barrels and that the amount of heat generated is proportional to the work done in turning a blunt boring tool. Rumford's observation of the proportionality between heat generated and work done lies at the foundation of thermodynamics. Another pioneer was the French military engineer Sadi Carnot, who introduced the concept of the heat-engine cycle and the principle of reversibility in 1824. Carnot's work concerned the limitations on the maximum amount of work that can be obtained from a steam engine operating with a high-temperature heat transfer as its driving force. Later that century, these ideas were developed by Rudolf Clausius, a German mathematician and physicist, into the first and second laws of thermodynamics, respectively.

The most important laws of thermodynamics are:

- The zeroth law of thermodynamics: When two systems are each in thermal equilibrium with a third system, the first two systems are in thermal equilibrium with each other. This property makes it meaningful to use thermometers as the "third system" and to define a temperature scale.
- The first law of thermodynamics, or the law of conservation of energy: The change in a system's internal energy is equal to the difference between heat added to the system from its surroundings and work done by the system on its surroundings.
- The second law of thermodynamics: Heat does not flow spontaneously from a colder region to a hotter region, or, equivalently, heat at a given temperature cannot be converted entirely into work. Consequently, the entropy of a closed system, or heat energy per unit temperature, increases over time toward some maximum value. Thus, all closed systems tend toward an equilibrium state in which entropy is at a maximum and no energy is available to do useful work. This asymmetry between forward and backward processes gives rise to what is known as the "arrow of time".

- The third law of thermodynamics: The entropy of a perfect crystal of an element in its most stable form tends to zero as the temperature approaches absolute zero. This allows an absolute scale for entropy to be established that, from a statistical point of view, determines the degree of randomness or disorder in a system.

Although thermodynamics developed rapidly during the 19th century in response to the need to optimize the performance of steam engines, the sweeping generality of the laws of thermodynamics makes them applicable to all physical and biological systems. In particular, the laws of thermodynamics give a complete description of all changes in the energy state of any system and its ability to perform useful work on its surroundings.

Fundamental Concepts

Temperature

The concept of temperature is fundamental to any discussion of thermodynamics, but its precise definition is not a simple matter. For example, a steel rod feels colder than a wooden rod at room temperature simply because steel is better at conducting heat away from the skin. It is therefore necessary to have an objective way of measuring temperature. In general, when two objects are brought into thermal contact, heat will flow between them until they come into equilibrium with each other. When the flow of heat stops, they are said to be at the same temperature. The zeroth law of thermodynamics formalizes this by asserting that if an object A is in simultaneous thermal equilibrium with two other objects B and C, then B and C will be in thermal equilibrium with each other if brought into thermal contact. Object A can then play the role of a thermometer through some change in its physical properties with temperature, such as its volume or its electrical resistance.

With the definition of equality of temperature in hand, it is possible to establish a temperature scale by assigning numerical values to certain easily reproducible fixed points. For example, in the Celsius (°C) temperature scale, the freezing point of pure water is arbitrarily assigned a temperature of 0 °C and the boiling point of water the value of 100 °C (in both cases at 1 standard atmosphere). In the Fahrenheit (°F) temperature scale, these same two points are assigned the values 32 °F and 212 °F, respectively. There are absolute temperature scales related to the second law of thermodynamics. The absolute scale related to the Celsius scale is called the Kelvin (K) scale, and that related to the Fahrenheit scale is called the Rankine (°R) scale. These scales are related by the equations K = °C + 273.15, °R = °F + 459.67, and °R = 1.8 K. Zero in both the Kelvin and Rankine scales is at absolute zero.

Work and Energy

Energy has a precise meaning in physics that does not always correspond to everyday language, and yet a precise definition is somewhat elusive. For example, a man pushing on a car may feel that he is doing a lot of work, but no work is actually done unless the car moves. The work done is then the product of the force applied by the man multiplied by the distance through which the car moves. If there is no friction and the surface is level, then the car, once set in motion, will continue rolling indefinitely with constant speed. The rolling car has something that a stationary car does not have—it has kinetic energy of motion equal to the work required to achieve that state of motion.

The introduction of the concept of energy in this way is of great value in mechanics because, in the absence of friction, energy is never lost from the system, although it can be converted from one form to another. For example, if a coasting car comes to a hill, it will roll some distance up the hill before coming to a temporary stop. At that moment its kinetic energy of motion has been converted into its potential energy of position, which is equal to the work required to lift the car through the same vertical distance. After coming to a stop, the car will then begin rolling back down the hill until it has completely recovered its kinetic energy of motion at the bottom. In the absence of friction, such systems are said to be conservative because at any given moment the total amount of energy (kinetic plus potential) remains equal to the initial work done to set the system in motion.

As the science of physics expanded to cover an ever-wider range of phenomena, it became necessary to include additional forms of energy in order to keep the total amount of energy constant for all closed systems (or to account for changes in total energy for open systems). For example, if work is done to accelerate charged particles, then some of the resultant energy will be stored in the form of electromagnetic fields and carried away from the system as radiation. In turn the electromagnetic energy can be picked up by a remote receiver (antenna) and converted back into an equivalent amount of work. With his theory of special relativity, Albert Einstein realized that energy (E) can also be stored as mass (m) and converted back into energy, as expressed by his famous equation $E = mc^2$, where c is the velocity of light. All of these systems are said to be conservative in the sense that energy can be freely converted from one form to another without limit. Each fundamental advance of physics into new realms has involved a similar extension to the list of the different forms of energy. In addition to preserving the first law of thermodynamics, also called the law of conservation of energy, each form of energy can be related back to an equivalent amount of work required to set the system into motion.

Thermodynamics encompasses all of these forms of energy, with the further addition of heat to the list of different kinds of energy. However, heat is fundamentally different from the others in that the conversion of work (or other forms of energy) into heat is not completely reversible, even in principle. In the example of the rolling car, some of the work done to set the car in motion is inevitably lost as heat due to friction, and the car eventually comes to a stop on a level surface. Even if all the generated heat were collected and stored in some fashion, it could never be converted entirely back into mechanical energy of motion. This fundamental limitation is expressed quantitatively by the second law of thermodynamics.

The role of friction in degrading the energy of mechanical systems may seem simple and obvious, but the quantitative connection between heat and work, as first discovered by Count Rumford, played a key role in understanding the operation of steam engines in the 19th century and similarly for all energy-conversion processes today.

Total Internal Energy

Although classical thermodynamics deals exclusively with the macroscopic properties of materials—such as temperature, pressure, and volume—thermal energy from the addition of heat can be understood at the microscopic level as an increase in the kinetic energy of motion of the molecules making up a substance. For example, gas molecules have translational kinetic energy that is proportional to the temperature of the gas: the molecules can rotate about their centre of mass, and the constituent atoms can vibrate with respect to each other (like masses connected by springs).

Additionally, chemical energy is stored in the bonds holding the molecules together, and weaker long-range interactions between the molecules involve yet more energy. The sum total of all these forms of energy constitutes the total internal energy of the substance in a given thermodynamic state. The total energy of a system includes its internal energy plus any other forms of energy, such as kinetic energy due to motion of the system as a whole (e.g., water flowing through a pipe) and gravitational potential energy due to its elevation.

Thermodynamic System

A thermodynamic system is defined as a definite quantity of matter or a region in space upon which attention is focussed in the analysis of a problem. We may want to study a quantity of matter contained with in a closed rigid walled chambers, or we may want to consider something such as gas pipeline through which the matter flows. The composition of the matter inside the system may be fixed or may change through chemical and nuclear reactions. A system may be arbitrarily defined. It becomes important when exchange of energy between the system and the everything else outside the system is considered. The judgement on the energetics of this exchange is very important.

Surroundings

Everything external to the system is surroundings. The system is distinguished from its surroundings by a specified boundary which may be at rest or in motion. The interactions between a system and its surroundings, which take place across the boundary, play an important role in thermodynamics. A system and its surroundings together comprise a universe.

Types of Thermodynamic System

On the basis of mass and energy transfer the thermodynamic system is divided into three types.

- Closed system
- Open system
- Isolated system

Closed system: A system in which the transfer of energy but not mass can takes place across the boundary is called closed system. The mass inside the closed system remains constant.

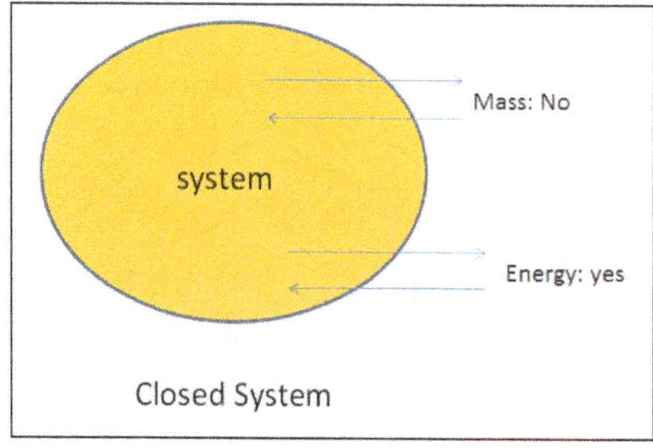

Closed System

For example: Boiling of water in a closed vessel. Since the water is boiled in closed vessel so the mass of water cannot escapes out of the boundary of the system but heat energy continuously entering and leaving the boundary of the vessel. It is an example of closed system.

Open system: A system in which the transfer of both mass and energy takes place is called an open system. This system is also known as control volume.

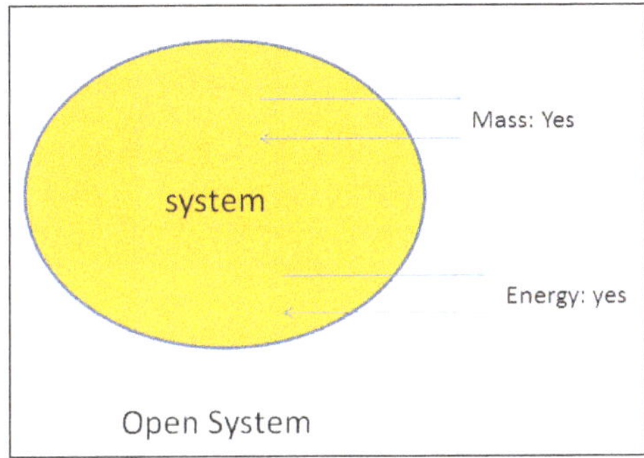

For example: Boiling of water in an open vessel is an example of open system because the water and heat energy both enters and leaves the boundary of the vessel.

Isolated system: A system in which the transfer of mass and energy cannot takes place is called an isolated system.

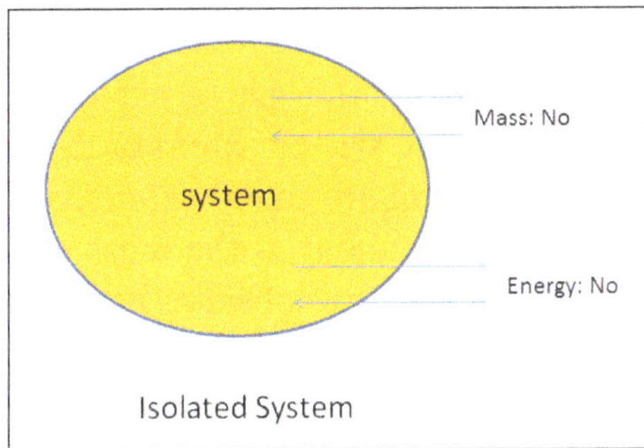

For example: Tea present in a thermos flask. In this the heat and the mass of the tea cannot cross the boundary of the thermos flask. Hence the thermos flak is an isolated system.

Property

To describe a system and predict its behaviour requires a knowledge of its properties and how those properties are related. Properties are macroscopic characteristics of a system such as mass, volume, energy, pressure and temperature to which numerical values can be assigned at a given time without knowledge of the past history of the system.

- The value of a property of a system is independent of the process or the path followed by the system in reaching a particular state.
- The change in the value of the property depends only on the initial and the final states.

The word state refers to the condition of a system as described by its properties.

Mathematically, if P is a property of the system, then the change in the property in going from the initial state 1 to the final state 2 is given by:

$$\int_1^2 dP = P_2 - P_1$$

If $P = P(x, y)$ then,

$$dP = \left(\frac{\partial P}{\partial x}\right)_y dx + \left(\frac{\partial P}{\partial y}\right)_x dy = adx + bdy$$

where,

$$a = \left(\frac{\partial P}{\partial x}\right)_y \text{ and } b = \left(\frac{\partial P}{\partial y}\right)_x$$

If $\left(\frac{\partial a}{\partial y}\right)_x = \left(\frac{\partial b}{\partial x}\right)_y$, then dP is said to be an exact differential, and P is a point function. A thermodynamic property is a point function and not a path function. Pressure, temperature, volume or molar volume are some of the quantities which satisfy these requirements.

Intensive and Extensive Properties

There are certain properties which depend on the size or extent of the system, and there are certain properties which are independent of the size or extent of the system. The properties like volume, which depend on the size of the system are called extensive properties. The properties, like temperature and pressure which are independent of the mass of the system are called intensive properties. The test for an intensive property is to observe how it is affected when a given system is combined with some fraction of exact replica of itself to create a new system differing only by size. Intensive properties are those which are unchanged by this process, whereas those properties whose values are increased or decreased in proportion to the enlargement or reduction of the system are called extensive properties.

Assume two identical systems S_1 and S_2 as shown in figure. Both the systems are in identical states.

Let S_3 be the combined system. Is the value of property for S_3 same as that for S_1 (and S_2)?

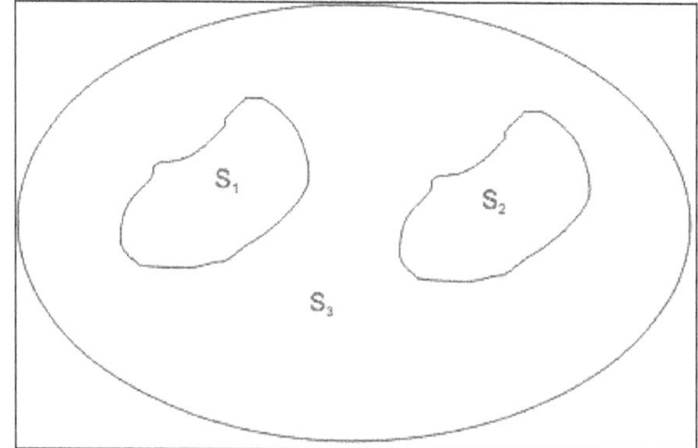

- If the answer is yes, then the property is intensive.
- If the answer is no, then the property is extensive.

The ratio of the extensive property to the mass is called the specific value of that property.

Specific volume, v = V/m = 1/ ρ (ρ is the density)

Specific internal energy, u = U/m

Similarly, the molar properties are defined as the ratios of the properties to the mole number (N) of the substance.

Molar volume = $\hat{v}$ = V/N

Molar internal energy = $\hat{u}$ = U/N

Thermodynamic State

A thermodynamic state is a set of values of properties of a thermodynamic system that must be specified to reproduce the system. The individual parameters are known as state variables, state parameters or thermodynamic variables. Once a sufficient set of thermodynamic variables have been specified, values of all other properties of the system are uniquely determined. The number of values required to specify the state depends on the system, and is not always known, but is often found from experimental evidence. Thermodynamics sets up an idealized formalism, in terms of idealized thermodynamic states, that can be summarized by a system of postulates of thermodynamics.

State Function

State function is a thermodynamic term that is used to name a property whose value does not depend on the path taken to reach that specific value. State functions are also known as point functions. A state function only depends on the current state of the thermodynamic system and its initial state (independent from the path). The state function of a thermodynamic system describes the equilibrium state of that system irrespective of how the system arrived at that state.

Examples of state functions:

- Mass
- Energy: Enthalpy, internal energy, Gibbs free energy, etc.
- Entropy
- Pressure
- Temperature
- Volume

- Chemical composition
- Altitude

A state function depends on three things: the property, initial value and final value. Enthalpy is a state function. It can be given as a mathematical expression as given below.

$$\int_{t_0}^{t_1} H(t)\,dt = H(t_1) - H(t_0)$$

In which, t_1 is the final state, t_0 is the initial state and h is the enthalpy of the system.

Path Functions

Path function is a thermodynamic term that is used to name a property whose value depends on the path taken to reach that specific value. In other words, a path function depends on the path taken to reach a final state from an initial state. Path function is also called a process function.

A path function gives different values for different paths. Hence path functions have variable values depending on the route. Therefore, when expressing the path function mathematically, multiple integrals and limits are required to integrate the path function.

Examples of Path Functions:

- Mechanical work
- Heat
- Arc length

The internal energy is given by the following equation:

$$\Delta U = q + w$$

In which ΔU is the change in internal energy, q is the heat and w is the mechanical work. The internal energy is a state function, but heat and work are path functions.

Difference between State Function and Path Function

State Function vs Path Function	
State function is a thermodynamic term that is used to name a property whose value does not depend on the path taken to reach that specific value.	Path function is a thermodynamic term that is used to name a property whose value depends on the path taken to reach that specific value.
Other Names	
State functions are also called point functions.	Path functions are also called process functions.
Process	
State functions do not depend on the path or process.	Path functions depend on the path or process.

Integration	
State function can be integrated using the initial and final values of the thermodynamic property of the system.	Path function requires multiple integrals and limits of integration to integrate the property.
Values	
The value of state function remains the same regardless of the number of steps.	The value of path function of a single step process is different from a multiple step process.
Examples	
State functions include entropy, enthalpy, mass, volume, temperature, etc.	Path functions include heat and mechanical work.

Thermodynamical Equilibrium

Let us suppose that there are two bodies at different temperatures, one hot and one cold. When these two bodies are brought in physical contact with each other, temperature of both the bodies will change. The hot body will tend to become colder while the cold body will tend to become hotter. Eventually both the bodies will achieve the same temperatures and they are said to be in thermodynamic equilibrium with each other. In an isolated system when there is no change in the macroscopic property of the system like entropy, internal energy etc, it is said to be in thermodynamic equilibrium. The state of the system which is in thermodynamic equilibrium is determined by intensive properties such as temperature, pressure, volume etc.

Whenever the system is in thermodynamic equilibrium, it tends to remain in this state infinitely and will not change spontaneously. Thus when the system is in thermodynamic equilibrium there won't be any spontaneous change in its macroscopic properties.

Conditions for Thermodynamic Equilibrium

The system is said to be in thermodynamic equilibrium if the conditions for following three equilibrium is satisfied:

- Mechanical equilibrium: When there are no unbalanced forces within the system and between the system and the surrounding, the system is said to be under mechanical equilibrium. The system is also said to be in mechanical equilibrium when the pressure throughout the system and between the system and surrounding is same. Whenever some unbalance forces exist within the system, they will get neutralized to attain the condition of equilibrium. Two systems are said to be in mechanical equilibrium with each other when their pressures are same.
- Chemical equilibrium: The system is said to be in chemical equilibrium when there are no chemical reactions going on within the system or there is no transfer of matter from one part of the system to other due to diffusion. Two systems are said to be in chemical equilibrium with each other when their chemical potentials are same.
- Thermal equilibrium: When the system is in mechanical and chemical equilibrium and

there is no spontaneous change in any of its properties, the system is said to be in thermal equilibrium. When the temperature of the system is uniform and not changing throughout the system and also in the surroundings, the system is said to be thermal equilibrium. Two systems are said to be thermal equilibrium with each other if their temperatures are same.

For the system to be thermodynamic equilibrium it is necessary that it should be under mechanical, chemical and thermal equilibrium. If any one of the condition are not fulfilled, the system is said to be in non-equilibrium.

References

- Thermodynamics: britannica.com, Retrieved 24 April, 2019
- System-Surroundings, mechanical-engineering: idc-online.com, Retrieved 24 April, 2019
- Thermodynamic-system-types-of-thermodynamic-system: mechanicalbooster.com, Retrieved 24 August, 2019
- Thermodynamic-state: definitions.net, Retrieved 15 May, 2019
- Difference-between-state-function-and-vs-path-function: differencebetween.com, Retrieved 29 June, 2019
- What-is-thermodynamic-equilibrium-part-one, thermodynamics- 4720: brighthubengineering.com, Retrieved 16 January, 2019

2

Laws of Thermodynamics

The laws of thermodynamics define entropy, temperature and energy which characterize thermodynamic systems and their equilibrium. It includes zeroth law, first law, second law and third law of thermodynamics. The topics elaborated in this chapter will help in gaining a better perspective about these laws of thermodynamics.

Zeroth Law of Thermodynamics

The Zeroth Law of Thermodynamics states that systems in thermal equilibrium are at the same temperature.

Systems are in thermal equilibrium if they do not transfer heat, even though they are in a position to do so, based on other factors. For example, food that's been in the refrigerator overnight is in thermal equilibrium with the air in the refrigerator: heat no longer flows from one source (the food) to the other source (the air) or back.

What the Zeroth Law of Thermodynamics means is that temperature is something worth measuring, because it indicates whether heat will move between objects. This will be true regardless of how the objects interact. Even if two objects don't touch, heat may still flow between them, such as by radiation (as from a heat lamp). However, according to the Zeroth Law of Thermodynamics, if the systems are in thermal equilibrium, no heat flow will take place.

There are more formal ways to state the Zeroth Law of Thermodynamics, which is commonly stated in the following manner:

Let A, B, and C be three systems. If A and C are in thermal equilibrium, and A and B are in thermal equilibrium, then B and C are in thermal equilibrium.

This statement is represented symbolically in. Temperature is not mentioned explicitly, but it's implied that temperature exists. Temperature is the quantity that is always the same for all systems in thermal equilibrium with one another.

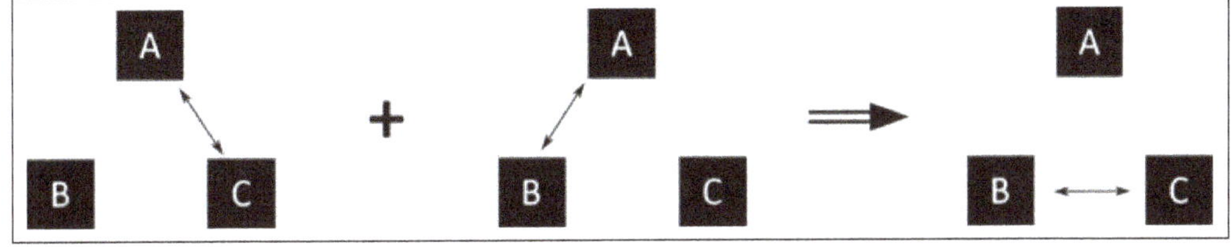

Zeroth Law of Thermodynamics: The double arrow represents thermal equilibrium between systems. If systems A and C are in equilibrium, and systems A and B are in equilibrium, then systems B and C are in equilibrium. The systems A, B, and C are at the same temperature.

Thermal Equilibrium

Thermal Equilibrium is a state in which neither matter nor energy has a net flow within a system or between systems. A single system can be in this state with itself or with other systems.

For example, a glass of ice water that has just been poured is not in thermodynamic equilibrium with itself. The ice and water are at different temperatures, so there is a net flow of energy from the warmer water to the cooler ice. After several minutes, once the ice has melted fully, the glass of chilled water is in thermodynamic equilibrium with itself: the (now melted) ice has warmed up and the water has cooled down.

On the other hand, the glass of chilled water and the room it is sitting in are two separate thermodynamic systems. They are connected because energy is allowed to flow between them. They are not in thermodynamic equilibrium because they are at different temperatures. Energy will flow from the room to the water, raising the temperature of the water a lot and decreasing the temperature of the room a little until the two are equal. The systems are now in thermodynamic equilibrium.

Zeroth Law as Equivalence Relation

A thermodynamic system is by definition in its own state of internal thermodynamic equilibrium, that is to say, there is no change in its observable state (i.e. macrostate) over time and no flows occur in it. One precise statement of the zeroth law is that the relation of thermal equilibrium is an equivalence relation on pairs of thermodynamic systems. In other words, the set of all systems each in its own state of internal thermodynamic equilibrium may be divided into subsets in which every system belongs to one and only one subset, and is in thermal equilibrium with every other member of that subset, and is not in thermal equilibrium with a member of any other subset. This means that a unique "tag" can be assigned to every system, and if the "tags" of two systems are the same, they are in thermal equilibrium with each other, and if different, they are not. This property is used to justify the use of empirical temperature as a tagging system. Empirical temperature provides further relations of thermally equilibrated systems, such as order and continuity with regard to "hotness" or "coldness", but these are not implied by the standard statement of the zeroth law.

If it is defined that a thermodynamic system is in thermal equilibrium with itself (i.e., thermal equilibrium is reflexive), then the zeroth law may be stated as follows:

> If a body C, be in thermal equilibrium with two other bodies, A and B, then A and B are in thermal equilibrium with one another.

This statement asserts that thermal equilibrium is a left-Euclidean relation between thermodynamic systems. If we also define that every thermodynamic system is in thermal equilibrium with itself, then thermal equilibrium is also a reflexive relation. Binary relations that are both reflexive and Euclidean are equivalence relations. Thus, again implicitly assuming reflexivity, the zeroth law is therefore often expressed as a right-Euclidean statement:

> If two systems are in thermal equilibrium with a third system, then they are in thermal equilibrium with each other.

One consequence of an equivalence relationship is that the equilibrium relationship is symmetric: If A is in thermal equilibrium with B, then B is in thermal equilibrium with A. Thus we may say that two systems are in thermal equilibrium with each other, or that they are in mutual equilibrium. Another consequence of equivalence is that thermal equilibrium is a transitive relationship and is occasionally expressed as such:

> If A is in thermal equilibrium with B and if B is in thermal equilibrium with C, then A is in thermal equilibrium with C.

A reflexive, transitive relationship does not guarantee an equivalence relationship. In order for the above statement to be true, both reflexivity and symmetry must be implicitly assumed.

It is the Euclidean relationships which apply directly to thermometry. An ideal thermometer is a thermometer which does not measurably change the state of the system it is measuring. Assuming that the unchanging reading of an ideal thermometer is a valid "tagging" system for the equivalence classes of a set of equilibrated thermodynamic systems, then if a thermometer gives the same reading for two systems, those two systems are in thermal equilibrium, and if we thermally connect the two systems, there will be no subsequent change in the state of either one. If the readings are different, then thermally connecting the two systems will cause a change in the states of both systems and when the change is complete, they will both yield the same thermometer reading. The zeroth law provides no information regarding this final reading.

Foundation of Temperature

The zeroth law establishes thermal equilibrium as an equivalence relationship. An equivalence relationship on a set (such as the set of all systems each in its own state of internal thermodynamic equilibrium) divides that set into a collection of distinct subsets ("disjoint subsets") where any member of the set is a member of one and only one such subset. In the case of the zeroth law, these subsets consist of systems which are in mutual equilibrium. This partitioning allows any member of the subset to be uniquely "tagged" with a label identifying the subset to which it belongs. Although the labeling may be quite arbitrary, temperature is just such a labeling process which uses the real number system for tagging. The zeroth law justifies the use of suitable thermodynamic systems as thermometers to provide such a labeling, which yield any number of possible empirical temperature scales, and justifies the use of the second law of thermodynamics to provide an absolute, or thermodynamic temperature scale. Such temperature scales bring additional continuity and ordering (i.e., "hot" and "cold") properties to the concept of temperature.

In the space of thermodynamic parameters, zones of constant temperature form a surface, that provides a natural order of nearby surfaces. One may therefore construct a global temperature function that provides a continuous ordering of states. The dimensionality of a surface of constant temperature is one less than the number of thermodynamic parameters, thus, for an ideal gas described with three thermodynamic parameters *P*, *V* and *N*, it is a two-dimensional surface.

For example, if two systems of ideal gases are in equilibrium, then $\frac{P_1V_1}{N_1} = \frac{P_2V_2}{N_2}$ where P_i is the pressure in the *i*th system, V_i is the volume, and N_i is the amount (in moles, or simply the number of atoms) of gas.

The surface $\frac{PV}{N}$ = constant defines surfaces of equal thermodynamic temperature, and one may label defining T so that $\frac{PV}{N} = RT$, where R is some constant. These systems can now be used as a thermometer to calibrate other systems. Such systems are known as "ideal gas thermometers".

In a sense, focused on in the zeroth law, there is only one kind of diathermal wall or one kind of heat, as expressed by Maxwell's dictum that "All heat is of the same kind". But in another sense, heat is transferred in different ranks, as expressed by Sommerfeld's dictum "Thermodynamics investigates the conditions that govern the transformation of heat into work. It teaches us to recognize temperature as the measure of the work-value of heat. Heat of higher temperature is richer, is capable of doing more work. Work may be regarded as heat of an infinitely high temperature, as unconditionally available heat." This is why temperature is the particular variable indicated by the zeroth law's statement of equivalence.

Physical Meaning of the usual Statement of the Zeroth Law

The usual statement of the zeroth law does not explicitly convey the full physical meaning that underlies it. The underlying physical meaning was perhaps first clarified by Maxwell in his 1871 textbook.

In Carathéodory's theory, it is postulated that there exist walls "permeable only to heat", though heat is not explicitly defined in that paper. This postulate is a physical postulate of existence. It does not, however, as worded just previously, say that there is only one kind of heat. This paper of Carathéodory states as proviso 4 of its account of such walls: "Whenever each of the systems S_1 and S_2 is made to reach equilibrium with a third system S_3 under identical conditions, systems S_1 and S_2 are in mutual equilibrium". It is the function of this statement in the paper, not there labeled as the zeroth law, to provide not only for the existence of transfer of energy other than by work or transfer of matter, but further to provide that such transfer is unique in the sense that there is only one kind of such wall, and one kind of such transfer. This is signaled in the postulate of this paper of Carathéodory that precisely one non-deformation variable is needed to complete the specification of a thermodynamic state, beyond the necessary deformation variables, which are not restricted in number. It is therefore not exactly clear what Carathéodory means when in the introduction of this paper he writes "It is possible to develop the whole theory without assuming the existence of heat, that is of a quantity that is of a different nature from the normal mechanical quantities."

Maxwell discusses at some length ideas which he summarizes by the words "All heat is of the same kind". Modern theorists sometimes express this idea by postulating the existence of a unique one-dimensional hotness manifold, into which every proper temperature scale has a monotonic mapping. This may be expressed by the statement that there is only one kind of temperature, regardless of the variety of scales in which it is expressed. Another modern expression of this idea is that "All diathermal walls are equivalent". This might also be expressed by saying that there is precisely one kind of non-mechanical, non-matter-transferring contact equilibrium between thermodynamic systems.

These ideas may be regarded as helping to clarify the physical meaning of the usual statement of the zeroth law of thermodynamics. It is the opinion of Lieb and Yngvason that the derivation from statistical mechanics of the law of entropy increase is a goal that has so far eluded the deepest thinkers. Thus the idea remains open to consideration that the existence of heat and temperature are needed as coherent primitive concepts for thermodynamics, as expressed, for example, by Maxwell and Planck. On the other hand, Planck clarified how the second law can be stated without reference to heat or temperature, by referring to the irreversible and universal nature of friction in natural thermodynamic processes.

Applications of Zeroth Law of Thermodynamics

The zeroth law of thermodynamics can be applied to compare temperatures of more than two objects. If the heat can pass through two objects without any external heating source, the two objects are said to be in thermal equilibrium. For instance, if the swimming pool you are in and your temperatures are similar, no heat flows from either sides apart from minute possibility. While, if you get into the pool cracking through the ice layer, you are not in the state of thermal equilibrium with the water. Please don't try this experiment. To check for thermal equilibrium mainly in case of frozen pools, you should use a thermometer.

The zeroth law frames an idea of temperature as an indicator of thermal equilibrium. The two given objects are in a state of equilibrium with the third object that gives you the required reading on a scale.

The zeroth law of thermodynamics is seen in many everyday situations:

- The thermometer may be the most well-known example of the zeroth law in action. For example, say the thermostat in your bedroom reads 67 degrees Fahrenheit. This means that the thermostat is in thermal equilibrium with your bedroom. However, because of the zeroth law of the thermodynamics, you can assume that both the room and other objects in the room (say, a clock hanging in the wall) are also at 67 degrees Fahrenheit.
- Similar to the above example, if you take a glass of ice water and a glass of hot water and place them on the kitchen countertop for a few hours, they will eventually reach thermal equilibrium with the room, with all 3 reaching the same temperature.
- If you place a package of meat in your freezer and leave it overnight, you assume that the meat has reached the same temperature as the freezer and the other items in the freezer.

First Law of Thermodynamics

The first law of thermodynamics is the physical law which states that the total energy of a system and its surroundings remain constant. The law is also known as the law of conservation of energy, which states energy can transform from one form into another, but can neither be created nor destroyed within an isolated system. Perpetual motion machines of the first kind are impossible, according to the first law of thermodynamics. In other words, it is not possible to construct an engine that will cycle and produce work continuously from nothing.

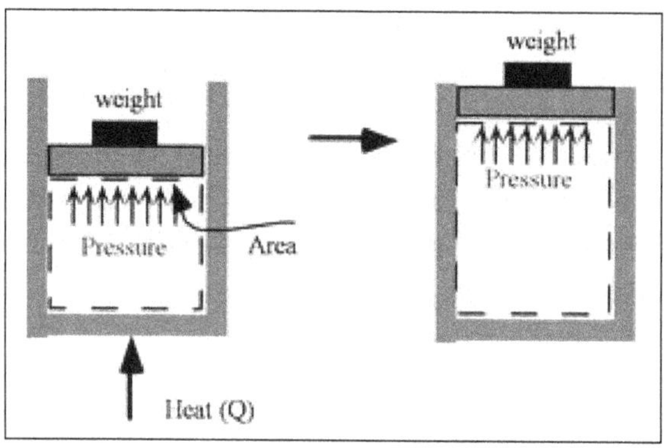

First Law of Thermodynamics Equation

The equation for the first law can be confusing because there are two different sign conventions in use.

In physics, particularly when discussing heat engines, the change in the energy of a system equals the heat flow in the system from the surroundings minus the work done by the system on the surroundings. The equation for the law may be written:

$$\Delta U = Q - W$$

Here, ΔU is the change in the internal energy of a closed system, Q is the heat supplied to the system, and W is the amount of work done by the system on the surroundings. This version of the law follows the sign convention of Clausius.

However, the IUPAC uses the sign convention proposed by Max Planck. Here, net energy transfer to a system is positive and net energy transfer from a system are negative. The equation then becomes:

$$\Delta U = Q + W$$

Cyclic Processes

The first law of thermodynamics for a closed system was expressed in two ways by Clausius. One way referred to cyclic processes and the inputs and outputs of the system, but did not refer to increments in the internal state of the system. The other way referred to an incremental change in the internal state of the system, and did not expect the process to be cyclic.

A cyclic process is one that can be repeated indefinitely often, returning the system to its initial state. Of particular interest for single cycle of a cyclic process are the net work done, and the net heat taken in (or 'consumed', in Clausius' statement), by the system.

In a cyclic process in which the system does net work on its surroundings, it is observed to be physically necessary not only that heat be taken into the system, but also, importantly, that some heat leave the system. The difference is the heat converted by the cycle into work. In each repetition of a cyclic process, the net work done by the system, measured in mechanical units, is proportional to the heat consumed, measured in calorimetric units.

The constant of proportionality is universal and independent of the system and in 1845 and 1847 was measured by James Joule, who described it as the mechanical equivalent of heat.

Sign Conventions

In a non-cyclic process, the change in the internal energy of a system is equal to net energy added as heat to the system minus the net work done by the system, both being measured in mechanical units. Taking ΔU as a change in internal energy, one writes:

$$\Delta U = Q - W \quad \text{(sign convention of Clausius)}$$

Where Q denotes the net quantity of heat supplied to the system by its surroundings and W denotes the net work done by the system. This sign convention is implicit in Clausius' statement of the law. It originated with the study of heat engines that produce useful work by consumption of heat.

Often nowadays, however, writers use the IUPAC convention by which the first law is formulated with work done on the system by its surroundings having a positive sign. With this now often used sign convention for work, the first law for a closed system may be written:

$$\Delta U = Q + W \quad \text{(sign convention of IUPAC).}$$

This convention follows physicists such as Max Planck, and considers all net energy transfers to the system as positive and all net energy transfers from the system as negative, irrespective of any use for the system as an engine or other device.

When a system expands in a fictive quasistatic process, the work done by the system on the environment is the product, P dV, of pressure, P, and volume change, dV, whereas the work done on the system is -P dV. Using either sign convention for work, the change in internal energy of the system is:

$$\mathrm{d}U = \delta Q - P\,\mathrm{d}V \quad \text{(quasi-static process),}$$

Where δQ denotes the infinitesimal amount of heat supplied to the system from its surroundings and δ denotes an inexact differential.

Work and heat are expressions of actual physical processes of supply or removal of energy, while the internal energy U is a mathematical abstraction that keeps account of the exchanges of energy that befall the system. Thus the term heat for Q means "that amount of energy added or removed by conduction of heat or by thermal radiation", rather than referring to a form of energy within the system. Likewise, the term work energy for W means "that amount of energy gained or lost as the result of work". Internal energy is a property of the system whereas work done and heat supplied are not. A significant result of this distinction is that a given internal energy change ΔU can be achieved by, in principle, many combinations of heat and work.

Various Statements of the Law for Closed Systems

The law is of great importance and generality and is consequently thought of from several points of view. Most careful textbook statements of the law express it for closed systems.

For the thermodynamics of closed systems, the distinction between transfers of energy as work and as heat is central.

There are two main ways of stating a law of thermodynamics, physically or mathematically. They should be logically coherent and consistent with one another.

An example of a physical statement is that of Planck:

> It is in no way possible, either by mechanical, thermal, chemical, or other devices, to obtain perpetual motion, i.e. it is impossible to construct an engine which will work in a cycle and produce continuous work, or kinetic energy, from nothing.

This physical statement is restricted neither to closed systems nor to systems with states that are strictly defined only for thermodynamic equilibrium; it has meaning also for open systems and for systems with states that are not in thermodynamic equilibrium.

An example of a mathematical statement is that of Crawford:

For a given system we let $\Delta E^{\text{kin}} =$ large-scale mechanical energy, $\Delta E^{\text{pot}} =$ large-scale potential energy, and $\Delta E^{\text{tot}} =$ total energy. The first two quantities are specifiable in terms of appropriate mechanical variables,

$$E^{\text{tot}} = E^{\text{kin}} + E^{\text{pot}} + U \ .$$

For any finite process, whether reversible or irreversible,

$$\Delta E^{\text{tot}} = \Delta E^{\text{kin}} + \Delta E^{\text{pot}} + \Delta U \ .$$

The first law in a form that involves the principle of conservation of energy more generally is,

$$\Delta E^{\text{tot}} = Q + W \ .$$

Here Q and W are heat and work added, with no restrictions as to whether the process is reversible, quasistatic, or irreversible.

This statement by Crawford, for W, uses the sign convention of IUPAC, not that of Clausius. Though it does not explicitly say so, this statement refers to closed systems, and to internal energy U defined for bodies in states of thermodynamic equilibrium, which possess well-defined temperatures.

The history of statements of the law for closed systems has two main periods, before and after the work of Bryan, of Carathéodory, and the approval of Carathéodory's work given by Born. The earlier traditional versions of the law for closed systems are nowadays often considered to be out of date.

Carathéodory's celebrated presentation of equilibrium thermodynamics refers to closed systems, which are allowed to contain several phases connected by internal walls of various kinds of impermeability and permeability (explicitly including walls that are permeable only to heat). Carathéodory's 1909 version of the first law of thermodynamics was stated in an axiom which refrained from defining or mentioning temperature or quantity of heat transferred. That axiom stated that the internal energy of a phase in equilibrium is a function of state, that the sum of the internal energies of the phases is the total internal energy of the system, and that the value of the total internal energy of the system is changed by the amount of work done adiabatically on it, considering work as a form of energy.

Such statements of the first law for closed systems assert the existence of internal energy as a function of state defined in terms of adiabatic work. Thus heat is not defined calorimetrically or as due to temperature difference. It is defined as a residual difference between change of internal energy and work done on the system, when that work does not account for the whole of the change of internal energy and the system is not adiabatically isolated.

The 1909 Carathéodory statement of the law in axiomatic form does not mention heat or temperature, but the equilibrium states to which it refers are explicitly defined by variable sets that necessarily include "non-deformation variables", such as pressures, which, within reasonable restrictions, can be rightly interpreted as empirical temperatures, and the walls connecting the phases of the system are explicitly defined as possibly impermeable to heat or permeable only to heat.

According to Münster, "A somewhat unsatisfactory aspect of Carathéodory's theory is that a consequence of the Second Law must be considered at this point [in the statement of the first law], i.e. that it is not always possible to reach any state 2 from any other state 1 by means of an adiabatic process." Münster instances that no adiabatic process can reduce the internal energy of a system at constant volume. Carathéodory's paper asserts that its statement of the first law corresponds exactly to Joule's experimental arrangement, regarded as an instance of adiabatic work. It does not point out that Joule's experimental arrangement performed essentially irreversible work, through friction of paddles in a liquid, or passage of electric current through a resistance inside the system, driven by motion of a coil and inductive heating, or by an external current source, which can access the system only by the passage of electrons, and so is not strictly adiabatic, because electrons are a form of matter, which cannot penetrate adiabatic walls. The paper goes on to base its main argument on the possibility of quasi-static adiabatic work, which is essentially reversible. The paper asserts that it will avoid reference to Carnot cycles, and then proceeds to base its argument on cycles of forward and backward quasi-static adiabatic stages, with isothermal stages of zero magnitude.

Sometimes the concept of internal energy is not made explicit in the statement.

Sometimes the existence of the internal energy is made explicit but work is not explicitly mentioned in the statement of the first postulate of thermodynamics. Heat supplied is then defined as the residual change in internal energy after work has been taken into account, in a non-adiabatic process.

The author then explains how heat is defined or measured by calorimetry, in terms of heat capacity, specific heat capacity, molar heat capacity, and temperature.

A respected text disregards the Carathéodory's exclusion of mention of heat from the statement of the first law for closed systems, and admits heat calorimetrically defined along with work and internal energy. Another respected text defines heat exchange as determined by temperature difference, but also mentions that the Born version is "completely rigorous". These versions follow the traditional approach that is now considered out of date, exemplified by that of Planck.

Evidence for the First Law of Thermodynamics for Closed Systems

The first law of thermodynamics for closed systems was originally induced from empirically observed evidence, including calorimetric evidence. It is nowadays, however, taken to provide the definition of heat via the law of conservation of energy and the definition of work in terms of changes in the external parameters of a system. The original discovery of the law was gradual over a period of perhaps half a century or more, and some early studies were in terms of cyclic processes.

The following is an account in terms of changes of state of a closed system through compound processes that are not necessarily cyclic. This account first considers processes for which the first law is easily verified because of their simplicity, namely adiabatic processes (in which there is no transfer as heat) and adynamic processes (in which there is no transfer as work).

Adiabatic Processes

In an adiabatic process, there is transfer of energy as work but not as heat. For all adiabatic process that takes a system from a given initial state to a given final state, irrespective of how the work is done, the respective eventual total quantities of energy transferred as work are one and the same, determined just by the given initial and final states. The work done on the system is defined and measured by changes in mechanical or quasi-mechanical variables external to the system. Physically, adiabatic transfer of energy as work requires the existence of adiabatic enclosures.

For instance, in Joule's experiment, the initial system is a tank of water with a paddle wheel inside. If we isolate the tank thermally, and move the paddle wheel with a pulley and a weight, we can relate the increase in temperature with the distance descended by the mass. Next, the system is returned to its initial state, isolated again, and the same amount of work is done on the tank using different devices (an electric motor, a chemical battery, a spring). In every case, the amount of work can be measured independently. The return to the initial state is not conducted by doing adiabatic work on the system. The evidence shows that the final state of the water (in particular, its temperature and volume) is the same in every case. It is irrelevant if the work is electrical, mechanical, chemical, or if done suddenly or slowly, as long as it is performed in an adiabatic way, that is to say, without heat transfer into or out of the system.

Evidence of this kind shows that to increase the temperature of the water in the tank, the qualitative kind of adiabatically performed work does not matter. No qualitative kind of adiabatic work has ever been observed to decrease the temperature of the water in the tank.

A change from one state to another, for example an increase of both temperature and volume, may be conducted in several stages, for example by externally supplied electrical work on a resistor in the body, and adiabatic expansion allowing the body to do work on the surroundings. It needs to be shown that the time order of the stages, and their relative magnitudes, does not affect the amount of adiabatic work that needs to be done for the change of state. According to one respected scholar: "Unfortunately, it does not seem that experiments of this kind have ever been carried out carefully. We must therefore admit that the statement which we have enunciated here, and which is equivalent to the first law of thermodynamics, is not well founded on direct experimental evidence." Another expression of this view is "no systematic precise experiments to verify this generalization directly have ever been attempted."

This kind of evidence, of independence of sequence of stages, combined with the evidence, of independence of qualitative kind of work, would show the existence of an important state variable that corresponds with adiabatic work, but not that such a state variable represented a conserved quantity. For the latter, another step of evidence is needed, which may be related to the concept of reversibility.

That important state variable was first recognized and denoted U by Clausius in 1850, but he did not then name it, and he defined it in terms not only of work but also of heat transfer in the same process. It was also independently recognized in 1850 by Rankine, who also denoted it U; and in 1851 by Kelvin who then called it "mechanical energy", and later "intrinsic energy". In 1865, after some hestitation, Clausius began calling his state function U "energy". In 1882 it was named as the *internal energy* by Helmholtz. If only adiabatic processes were of interest, and heat could be ignored, the concept of internal energy would hardly arise or be needed. The relevant physics would be largely covered by the concept of potential energy, as was intended in the 1847 paper of Helmholtz on the principle of conservation of energy, though that did not deal with forces that cannot be described by a potential, and thus did not fully justify the principle. Moreover, that paper was critical of the early work of Joule that had by then been performed. A great merit of the internal energy concept is that it frees thermodynamics from a restriction to cyclic processes, and allows a treatment in terms of thermodynamic states.

In an adiabatic process, adiabatic work takes the system either from a reference state O with internal energy U(O) to an arbitrary one A with internal energy U(A), or from the state A to the state O:

$$U(A) = U(O) - W_{O\to A}^{\text{adiabatic}} \text{ or } U(O) = U(A) - W_{A\to O}^{\text{adiabatic}}.$$

Except under the special, and strictly speaking, fictional, condition of reversibility, only one of the processes adiabatic, $O \to A$ or adiabatic, $A \to O$ is empirically feasible by a simple application of externally supplied work. The reason for this is given as the second law of thermodynamics.

The fact of such irreversibility may be dealt with in two main ways, according to different points of view:

- Since the work of Bryan, the most accepted way to deal with it nowadays, followed by Carathéodory, is to rely on the previously established concept of quasi-static processes, as follows. Actual physical processes of transfer of energy as work are always at least to some degree irreversible. The irreversibility is often due to mechanisms known as dissipative, that transform bulk kinetic energy into internal energy. Examples are friction and viscosity. If the process is performed more slowly, the frictional or viscous dissipation is less. In the limit of infinitely slow performance, the dissipation tends to zero and then the limiting process, though fictional rather than actual, is notionally reversible, and is called quasi-static. Throughout the course of the fictional limiting quasi-static process, the internal intensive variables of the system are equal to the external intensive variables, those that describe the reactive forces exerted by the surroundings. This can be taken to justify the formula:

$$W_{A\to O}^{\text{adiabatic, quasi-static}} = -W_{O\to A}^{\text{adiabatic, quasi-static}}$$

- Another way to deal with it is to allow that experiments with processes of heat transfer to or from the system may be used to justify the formula above. Moreover, it deals to some extent with the problem of lack of direct experimental evidence that the time order of stages of a process does not matter in the determination of internal energy. This way does not provide theoretical purity in terms of adiabatic work processes, but is empirically feasible, and is in accord with experiments actually done, such as the Joule experiments, and with older traditions.

The formula allows that to go by processes of quasi-static adiabatic work from the state A to the state B we can take a path that goes through the reference state O, since the quasi-static adiabatic work is independent of the path,

$$-W_{A\to B}^{\text{adiabatic, quasi-static}} = -W_{A\to O}^{\text{adiabatic, quasi-static}} - W_{O\to B}^{\text{adiabatic, quasi-static}}$$
$$= W_{O\to A}^{\text{adiabatic, quasi-static}} - W_{O\to B}^{\text{adiabatic, quasi-static}} = -U(A) + U(B) = \Delta U$$

This kind of empirical evidence, coupled with theory of this kind, largely justifies the following statement:

> "For all adiabatic processes between two specified states of a closed system of any nature, the net work done is the same regardless the details of the process, and determines a state function called internal energy, U".

Adynamic Processes

A complementary observable aspect of the first law is about heat transfer. Adynamic transfer of energy as heat can be measured empirically by changes in the surroundings of the system of interest by calorimetry. This again requires the existence of adiabatic enclosure of the entire process, system and surroundings, though the separating wall between the surroundings and the system is thermally conductive or radiatively permeable, not adiabatic. A calorimeter can rely on measurement of sensible heat, which requires the existence of thermometers and measurement of temperature change in bodies of known sensible heat capacity under specified conditions; or it can rely on the measurement of latent heat, through measurement of masses of material that change phase, at temperatures fixed by the occurrence of phase changes under specified conditions in bodies of known latent heat of phase change. The calorimeter can be calibrated by adiabatically doing externally determined work on it. The most accurate method is by passing an electric current from outside through a resistance inside the calorimeter. The calibration allows comparison of calorimetric measurement of quantity of heat transferred with quantity of energy transferred as work. "The most common device for measuring ΔU is an adiabatic bomb calorimeter." "Calorimetry is widely used in present day laboratories." According to one opinion, "Most thermodynamic data come from calorimetry." According to another opinion, "The most common method of measuring "heat" is with a calorimeter."

When the system evolves with transfer of energy as heat, without energy being transferred as work, in an adynamic process, the heat transferred to the system is equal to the increase in its internal energy:

$$Q_{A\to B}^{\text{adynamic}} = \Delta U$$

General Case for Reversible Processes

Heat transfer is practically reversible when it is driven by practically negligibly small temperature gradients. Work transfer is practically reversible when it occurs so slowly that there are no frictional effects within the system; frictional effects outside the system should also be zero if the process is to be globally reversible. For a particular reversible process in general, the work done reversibly on the system, $W_{A\to B}^{\text{path } P_0,\text{reversible}}$, and the heat transferred reversibly to the system, $Q_{A\to B}^{\text{path } P_0,\text{reversible}}$ are not required to occur respectively adiabatically or adynamically, but they must belong to the same particular process defined by its particular reversible path, P_0, through the space of thermodynamic states. Then the work and heat transfers can occur and be calculated simultaneously.

Putting the two complementary aspects together, the first law for a particular reversible process can be written,

$$-W_{A\to B}^{\text{path } P_0,\text{reversible}} + Q_{A\to B}^{\text{path } P_0,\text{reversible}} = \Delta U$$

This combined statement is the expression the first law of thermodynamics for reversible processes for closed systems.

In particular, if no work is done on a thermally isolated closed system we have,

$$\Delta U = 0.$$

This is one aspect of the law of conservation of energy and can be stated:

> The internal energy of an isolated system remains constant.

General Case for Irreversible Processes

If, in a process of change of state of a closed system, the energy transfer is not under a practically zero temperature gradient and practically frictionless, then the process is irreversible. Then the heat and work transfers may be difficult to calculate, and irreversible thermodynamics is called for. Nevertheless, the first law still holds and provides a check on the measurements and calculations of the work done irreversibly on the system, $W_{A\to B}^{\text{path } P_1,\text{ irreversible}}$, and the heat transferred irreversibly to the system, $Q_{A\to B}^{\text{path } P_1,\text{ irreversible}}$, which belong to the same particular process defined by its particular irreversible path, P_1, through the space of thermodynamic states.

$$-W_{A\to B}^{\text{path } P_1,\text{ irreversible}} + Q_{A\to B}^{\text{path } P_1,\text{ irreversible}} = \Delta U.$$

This means that the internal energy U is a function of state and that the internal energy change ΔU between two states is a function only of the two states.

Weight of Evidence for the Law

The first law of thermodynamics is so general that its predictions cannot all be directly tested. In many properly conducted experiments it has been precisely supported, and never violated. Indeed, within its scope of applicability, the law is so reliably established, that, nowadays, rather than experiment being considered as testing the accuracy of the law, it is more practical and realistic to think of the law as testing the accuracy of experiment. An experimental result that seems to violate the law may be assumed to be inaccurate or wrongly conceived, for example due to failure to account for an important physical factor. Thus, some may regard it as a principle more abstract than a law.

State Functional Formulation for Infinitesimal Processes

When the heat and work transfers in the equations are infinitesimal in magnitude, they are often denoted by δ, rather than exact differentials denoted by d, as a reminder that heat and work do not describe the *state* of any system. The integral of an inexact differential depends upon the particular path taken through the space of thermodynamic parameters while the integral of an exact differential depends only upon the initial and final states. If the initial and final states are the same, then the integral of an inexact differential may or may not be zero, but the integral of an exact differential is always zero. The path taken by a thermodynamic system through a chemical or physical change is known as a thermodynamic process.

The first law for a closed homogeneous system may be stated in terms that include concepts that are established in the second law. The internal energy U may then be expressed as a function of the system's defining state variables S, entropy, and V, volume: $U = U(S, V)$. In these terms, T, the system's temperature, and P, its pressure, are partial derivatives of U with respect to S and V. These variables are important throughout thermodynamics, though not necessary for the statement of the first law. Rigorously, they are defined only when the system is in its own state of internal thermodynamic equilibrium. For some purposes, the concepts provide good approximations for scenarios sufficiently near to the system's internal thermodynamic equilibrium.

The first law requires that:

$$dU = \delta Q - \delta W \qquad \text{(closed system, general process, quasi-static or irreversible).}$$

Then, for the fictive case of a reversible process, dU can be written in terms of exact differentials. One may imagine reversible changes, such that there is at each instant negligible departure from thermodynamic equilibrium within the system. This excludes isochoric work. Then, mechanical work is given by $\delta W = -P\,dV$ and the quantity of heat added can be expressed as $\delta Q = T\,dS$. For these conditions:

$$dU = TdS - PdV \qquad \text{(closed system, reversible process).}$$

While this has been shown here for reversible changes, it is valid in general, as U can be considered as a thermodynamic state function of the defining state variables S and V:

$$dU = TdS - PdV \quad \text{(closed system, general process, quasi-static or irreversible).}$$

Equation $dU = TdS - PdV$ is known as the fundamental thermodynamic relation for a closed system in the energy representation, for which the defining state variables are S and V, with respect to which T and P are partial derivatives of U. It is only in the fictive reversible case, when isochoric work is excluded, that the work done and heat transferred are given by $-PdV$ and TdS.

In the case of a closed system in which the particles of the system are of different types and, because chemical reactions may occur, their respective numbers are not necessarily constant, the fundamental thermodynamic relation for dU becomes:

$$dU = TdS - PdV + \sum_i \mu_i dN_i.$$

where dN_i is the (small) increase in number of type-i particles in the reaction, and μ_i is known as the chemical potential of the type-i particles in the system. If dN_i is expressed in mol then μ_i is expressed in J/mol. If the system has more external mechanical variables than just the volume that can change, the fundamental thermodynamic relation further generalizes to:

$$dU = TdS - \sum_i X_i dx_i + \sum_j \mu_j dN_j.$$

Here the X_i are the generalized forces corresponding to the external variables x_i. The parameters X_i are independent of the size of the system and are called intensive parameters and the x_i are proportional to the size and called extensive parameters.

For an open system, there can be transfers of particles as well as energy into or out of the system during a process. For this case, the first law of thermodynamics still holds, in the form that the internal energy is a function of state and the change of internal energy in a process is a function only of its initial and final states.

A useful idea from mechanics is that the energy gained by a particle is equal to the force applied to the particle multiplied by the displacement of the particle while that force is applied. Now consider the first law without the heating term: dU = −PdV. The pressure P can be viewed as a force (and in fact has units of force per unit area) while dV is the displacement (with units of distance times area). We may say, with respect to this work term, that a pressure difference forces a transfer of volume, and that the product of the two (work) is the amount of energy transferred out of the system as a result of the process. If one were to make this term negative then this would be the work done on the system.

It is useful to view the TdS term in the same light: here the temperature is known as a "generalized" force (rather than an actual mechanical force) and the entropy is a generalized displacement.

Similarly, a difference in chemical potential between groups of particles in the system drives a chemical reaction that changes the numbers of particles, and the corresponding product is the amount of chemical potential energy transformed in process. For example, consider a system consisting of two phases: liquid water and water vapor. There is a generalized "force" of evaporation that drives water molecules out of the liquid. There is a generalized "force" of condensation that drives vapor molecules out of the vapor. Only when these two "forces" (or chemical potentials) are equal is there equilibrium, and the net rate of transfer zero.

The two thermodynamic parameters that form a generalized force-displacement pair are called "conjugate variables". The two most familiar pairs are, of course, pressure-volume, and temperature-entropy.

Spatially Inhomogeneous Systems

Classical thermodynamics is initially focused on closed homogeneous systems (e.g. Planck 1897/1903), which might be regarded as 'zero-dimensional' in the sense that they have no spatial variation. But it is desired to study also systems with distinct internal motion and spatial inhomogeneity. For such systems, the principle of conservation of energy is expressed in terms not only of internal energy as defined for homogeneous systems, but also in terms of kinetic energy and potential energies of parts of the inhomogeneous system with respect to each other and with respect to long-range external forces. How the total energy of a system is allocated between these three more specific kinds of energy varies according to the purposes of different writers; this is because these components of energy are to some extent mathematical artefacts rather than actually measured physical quantities. For any closed homogeneous component of an inhomogeneous closed system, if E denotes the total energy of that component system, one may write:

$$E = E^{\text{kin}} + E^{\text{pot}} + U$$

Where E^{kin} and E^{pot} denote respectively the total kinetic energy and the total potential energy of the component closed homogeneous system, and U denotes its internal energy.

Potential energy can be exchanged with the surroundings of the system when the surroundings impose a force field, such as gravitational or electromagnetic, on the system.

A compound system consisting of two interacting closed homogeneous component subsystems has a potential energy of interaction E_{12}^{pot} between the subsystems. Thus, in an obvious notation, one may write:

$$E = E_1^{\text{kin}} + E_1^{\text{pot}} + U_1 + E_2^{\text{kin}} + E_2^{\text{pot}} + U_2 + E_{12}^{\text{pot}}$$

The quantity E_{12}^{pot} in general lacks an assignment to either subsystem in a way that is not arbitrary, and this stands in the way of a general non-arbitrary definition of transfer of energy as work. On occasions, authors make their various respective arbitrary assignments.

The distinction between internal and kinetic energy is hard to make in the presence of turbulent motion within the system, as friction gradually dissipates macroscopic kinetic energy of localised bulk flow into molecular random motion of molecules that is classified as internal energy. The rate of dissipation by friction of kinetic energy of localised bulk flow into internal energy, whether in turbulent or in streamlined flow, is an important quantity in non-equilibrium thermodynamics. This is a serious difficulty for attempts to define entropy for time-varying spatially inhomogeneous systems.

First Law of Thermodynamics for Open Systems

For the first law of thermodynamics, there is no trivial passage of physical conception from the closed system view to an open system view. For closed systems, the concepts of an adiabatic enclosure and of an adiabatic wall are fundamental. Matter and internal energy cannot permeate or penetrate such a wall. For an open system, there is a wall that allows penetration by matter. In general, matter in diffusive motion carries with it some internal energy, and some microscopic potential energy changes accompany the motion. An open system is not adiabatically enclosed.

There are some cases in which a process for an open system can, for particular purposes, be considered as if it were for a closed system. In an open system, by definition hypothetically or potentially, matter can pass between the system and its surroundings. But when, in a particular case, the process of interest involves only hypothetical or potential but no actual passage of matter, the process can be considered as if it were for a closed system.

Internal Energy for an Open System

Since the revised and more rigorous definition of the internal energy of a closed system rests upon the possibility of processes by which adiabatic work takes the system from one state to another, this leaves a problem for the definition of internal energy for an open system, for which adiabatic work is not in general possible. According to Max Born, the transfer of matter and energy across an open connection "cannot be reduced to mechanics". In contrast to the case of closed systems, for open systems, in the presence of diffusion, there is no unconstrained and unconditional physical distinction between convective transfer of internal energy by bulk flow of matter, the transfer of internal energy without transfer of matter (usually called heat conduction and work transfer), and change of various potential energies. The older traditional way and the conceptually revised (Carathéodory) way agree that there is no physically unique definition of heat and work transfer processes between open systems.

In particular, between two otherwise isolated open systems an adiabatic wall is by definition impossible. This problem is solved by recourse to the principle of conservation of energy. This principle allows a composite isolated system to be derived from two other component non-interacting isolated systems, in such a way that the total energy of the composite isolated system is equal to the sum of the total energies of the two component isolated systems. Two previously isolated systems can be subjected to the thermodynamic operation of placement between them of a wall permeable to matter and energy, followed by a time for establishment of a new thermodynamic state of internal equilibrium in the new single unpartitioned system. The internal energies of the initial two systems and of the final new system, considered respectively as closed systems as above, can be measured. Then the law of conservation of energy requires that,

$$\Delta U_s + \Delta U_o = 0$$

Where ΔU_s and ΔU_o denote the changes in internal energy of the system and of its surroundings respectively. This is a statement of the first law of thermodynamics for a transfer between two otherwise isolated open systems, that fits well with the conceptually revised and rigorous statement of the law.

For the thermodynamic operation of adding two systems with internal energies U_1 and U_2, to produce a new system with internal energy U, one may write $U = U_1 + U_2$; the reference states for U, U_1 and U_2 should be specified accordingly, maintaining also that the internal energy of a system be proportional to its mass, so that the internal energies are extensive variables.

There is a sense in which this kind of additivity expresses a fundamental postulate that goes beyond the simplest ideas of classical closed system thermodynamics; the extensivity of some variables is not obvious, and needs explicit expression; indeed one author goes so far as to say that it could be recognized as a fourth law of thermodynamics, though this is not repeated by other authors.

Also of course

$$\Delta N_s + \Delta N_o = 0$$

Where ΔN_s and ΔN_o denote the changes in mole number of a component substance of the system and of its surroundings respectively. This is a statement of the law of conservation of mass.

Process of Transfer of Matter between an Open System and its Surroundings

A system connected to its surroundings only through contact by a single permeable wall, but otherwise isolated, is an open system. If it is initially in a state of contact equilibrium with a surrounding subsystem, a thermodynamic process of transfer of matter can be made to occur between them if the surrounding subsystem is subjected to some thermodynamic operation, for example, removal of a partition between it and some further surrounding subsystem. The removal of the partition in the surroundings initiates a process of exchange between the system and its contiguous surrounding subsystem.

An example is evaporation. One may consider an open system consisting of a collection of liquid, enclosed except where it is allowed to evaporate into or to receive condensate from its vapor above it, which may be considered as its contiguous surrounding subsystem, and subject to control of its volume and temperature.

A thermodynamic process might be initiated by a thermodynamic operation in the surroundings, that mechanically increases in the controlled volume of the vapor. Some mechanical work will be done within the surroundings by the vapor, but also some of the parent liquid will evaporate and enter the vapor collection which is the contiguous surrounding subsystem. Some internal energy will accompany the vapor that leaves the system, but it will not make sense to try to uniquely identify part of that internal energy as heat and part of it as work. Consequently, the energy transfer that accompanies the transfer of matter between the system and its surrounding subsystem cannot be uniquely split into heat and work transfers to or from the open system. The component of total energy transfer that accompanies the transfer of vapor into the surrounding subsystem is customarily called 'latent heat of evaporation', but this use of the word heat is a quirk of customary historical language, not in strict compliance with the thermodynamic definition of transfer of energy as heat. In this example, kinetic energy of bulk flow and potential energy with respect to long-range external forces such as gravity are both considered to be zero. The first law of thermodynamics refers to the change of internal energy of the open system, between its initial and final states of internal equilibrium.

Open System with Multiple Contacts

An open system can be in contact equilibrium with several other systems at once.

This includes cases in which there is contact equilibrium between the system, and several subsystems in its surroundings, including separate connections with subsystems through walls that are permeable to the transfer of matter and internal energy as heat and allowing friction of passage of the transferred matter, but immovable, and separate connections through adiabatic walls with others, and separate connections through diathermic walls impermeable to matter with yet others. Because there are physically separate connections that are permeable to energy but impermeable to matter, between the system and its surroundings, energy transfers between them can occur with definite heat and work characters. Conceptually essential here is that the internal energy transferred with the transfer of matter is measured by a variable that is mathematically independent of the variables that measure heat and work.

With such independence of variables, the total increase of internal energy in the process is then determined as the sum of the internal energy transferred from the surroundings with the transfer of matter through the walls that are permeable to it, and of the internal energy transferred to the system as heat through the diathermic walls, and of the energy transferred to the system as work through the adiabatic walls, including the energy transferred to the system by long-range forces. These simultaneously transferred quantities of energy are defined by events in the surroundings of the system. Because the internal energy transferred with matter is not in general uniquely resolvable into heat and work components, the total energy transfer cannot in general be uniquely resolved into heat and work components. Under these conditions, the following formula can describe the process in terms of externally defined thermodynamic variables, as a statement of the first law of thermodynamics:

$$\Delta U_0 = Q - W - \sum_{i=1}^{m} \Delta U_i$$

(suitably defined surrounding subsystems, general process, quasi-static or irreversible)

where ΔU_0 denotes the change of internal energy of the system, and ΔU_i denotes the change of internal energy of the ith of the m surrounding subsystems that are in open contact with the system, due to transfer between the system and that ith surrounding subsystem, and Q denotes the internal energy transferred as heat from the heat reservoir of the surroundings to the system, and W denotes the energy transferred from the system to the surrounding subsystems that are in adiabatic connection with it. The case of a wall that is permeable to matter and can move so as to allow transfer of energy as work is not considered here.

Combination of First and Second Laws

If the system is described by the energetic fundamental equation, $U_0 = U_0(S, V, N_j)$, and if the process can be described in the quasi-static formalism, in terms of the internal state variables of the system, then the process can also be described by a combination of the first and second laws of thermodynamics, by the formula:

$$\mathrm{d}U_0 = T\,\mathrm{d}S - P\,\mathrm{d}V + \sum_{j=1}^{n} \mu_j\,\mathrm{d}N_j$$

Where there are n chemical constituents of the system and permeably connected surrounding subsystems, and where T, S, P, V, N_j, and μ_j, are defined as above.

For a general natural process, there is no immediate term-wise correspondence between equations $\Delta U_0 = Q - W - \sum_{i=1}^{m} \Delta U_i$ and $\mathrm{d}U_0 = T\,\mathrm{d}S - P\,\mathrm{d}V + \sum_{j=1}^{n} \mu_j\,\mathrm{d}N_j$, because they describe the process in different conceptual frames.

Nevertheless, a conditional correspondence exists. There are three relevant kinds of wall here: purely diathermal, adiabatic, and permeable to matter. If two of those kinds of wall are sealed off, leaving only one that permits transfers of energy, as work, as heat, or with matter, then the remaining permitted terms correspond precisely. If two of the kinds of wall are left unsealed, then energy transfer can be shared between them, so that the two remaining permitted terms do not correspond precisely.

For the special fictive case of quasi-static transfers, there is a simple correspondence. For this, it is supposed that the system has multiple areas of contact with its surroundings. There are pistons that allow adiabatic work, purely diathermal walls, and open connections with surrounding subsystems of completely controllable chemical potential (or equivalent controls for charged species). Then, for a suitable fictive quasi-static transfer, one can write,

$$\delta Q = T\,\mathrm{d}S \text{ and } \delta W = P\,\mathrm{d}V \text{ (suitably defined surrounding subsystems, quasi-static transfers of energy).}$$

For fictive quasi-static transfers for which the chemical potentials in the connected surrounding subsystems are suitably controlled, these can be put into equation $\mathrm{d}U_0 = T\,\mathrm{d}S - P\,\mathrm{d}V + \sum_{j=1}^{n} \mu_j\,\mathrm{d}N_j$ to yield:

$$\mathrm{d}U_0 = \delta Q - \delta W + \sum_{j=1}^{n} \mu_j\,\mathrm{d}N_j \text{ (suitably defined surrounding subsystems, quasi-static transfers).}$$

The reference does not actually write equation $\mathrm{d}U_0 = \delta Q - \delta W + \sum_{j=1}^{n} \mu_j\,\mathrm{d}N_j$, but what it does write is fully compatible with it. Another helpful account is given by Tschoegl.

There are several other accounts of this, in apparent mutual conflict.

Non-equilibrium Transfers

The transfer of energy between an open system and a single contiguous subsystem of its surroundings is considered also in non-equilibrium thermodynamics. The problem of definition arises also in this case. It may be allowed that the wall between the system and the subsystem is not only permeable to matter and to internal energy, but also may be movable so as to allow work to be done when the two systems have different pressures. In this case, the transfer of energy as heat is not defined.

Methods for study of non-equilibrium processes mostly deal with spatially continuous flow systems. In this case, the open connection between system and surroundings is usually taken to fully surround the system, so that there are no separate connections impermeable to matter but permeable to heat. Except for the special case mentioned above when there is no actual transfer of matter, which can be treated as if for a closed system, in strictly defined thermodynamic terms, it follows that transfer of energy as heat is not defined. In this sense, there is no such thing as 'heat flow' for a continuous-flow open system. Properly, for closed systems, one speaks of transfer of internal energy as heat, but in general, for open systems, one can speak safely only of transfer of internal energy. A factor here is that there are often cross-effects between distinct transfers, for example that transfer of one substance may cause transfer of another even when the latter has zero chemical potential gradient.

Usually transfer between a system and its surroundings applies to transfer of a state variable, and obeys a balance law, that the amount lost by the donor system is equal to the amount gained by the receptor system. Heat is not a state variable. For his 1947 definition of "heat transfer" for discrete open systems, the author Prigogine carefully explains at some length that his definition of it does not obey a balance law. He describes this as paradoxical.

The situation is clarified by Gyarmati, who shows that his definition of "heat transfer", for continuous-flow systems, really refers not specifically to heat, but rather to transfer of internal energy, as follows. He considers a conceptual small cell in a situation of continuous-flow as a system defined in the so-called Lagrangian way, moving with the local center of mass. The flow of matter across the boundary is zero when considered as a flow of total mass. Nevertheless, if the material constitution is of several chemically distinct components that can diffuse with respect to one another, the system is considered to be open, the diffusive flows of the components being defined with respect to the center of mass of the system, and balancing one another as to mass transfer. Still there can be a distinction between bulk flow of internal energy and diffusive flow of internal energy in this case, because the internal energy density does not have to be constant per unit mass of material, and allowing for non-conservation of internal energy because of local conversion of kinetic energy of bulk flow to internal energy by viscosity.

Heat Transfer

Heat is defined as the form of energy that is transferred between two systems by virtue of a temperature difference.

There cannot be any heat transfer between two systems that are at the same temperature.

It is the thermal (internal) energy that can be stored in a system. Heat is a form of energy in transition and as a result can only be identified at the system boundary.

Heat has energy units kJ (or BTU). Rate of heat transfer is the amount of heat transferred per unit time.

Heat is a directional (or vector) quantity; thus, it has magnitude, direction and point of action.

Notation:

- Q (kJ) amount of heat transfer.
- Q° (kW) rate of heat transfer (power).
- q (kJ/kg) - heat transfer per unit mass.
- q° (kW/kg) - power per unit mass.

Sign convention: Heat Transfer to a system is positive, and heat transfer from a system is negative. It means any heat transfer that increases the energy of a system is positive, and heat transfer that decreases the energy of a system is negative.

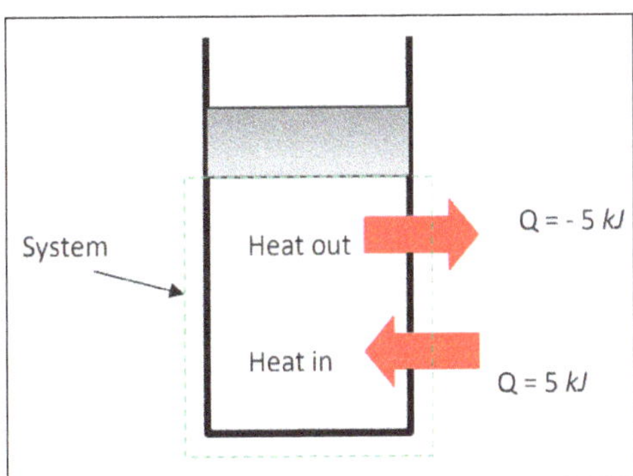

Sign convention: positive if to the system, negative if from the system.

Modes of Heat Transfer

Heat can be transferred in three different modes conduction, convection, and radiation. All modes of heat transfer require the existence of a temperature difference.

Conduction is the transfer of energy from the more energetic particles to the adjacent less energetic particles as a result of interactions between particles.

In solids, conduction is due to the combination of vibrations of the molecules in a lattice and the energy transport by free electrons.

Convection is the mode of energy transfer between a solid surface and the adjacent liquid or gas which is in motion, and it involves the combined effects of conduction and fluid motion (advection).

Convection is called forced if the fluid is forced to flow by external means such as a fan or a pump. It is called free or natural if the fluid motion is caused by buoyancy forces that are induced by density differences due to the temperature variation in a fluid.

Radiation is the energy emitted by matter in the form of electromagnetic waves (or photons) as a result of the changes in the electronic configurations of the atoms or molecules.

Work

Work is the energy interaction between a system and its surroundings. More specifically, work is the energy transfer associated with force acting through a distance.

Notation:

- W (kJ) amount of work transfer
- W° (kW) power
- w (kJ/kg) - work per unit mass
- w° (kW/kg) - power per unit mass

Sign convention are work done by a system is positive, and the work done on a system is negative.

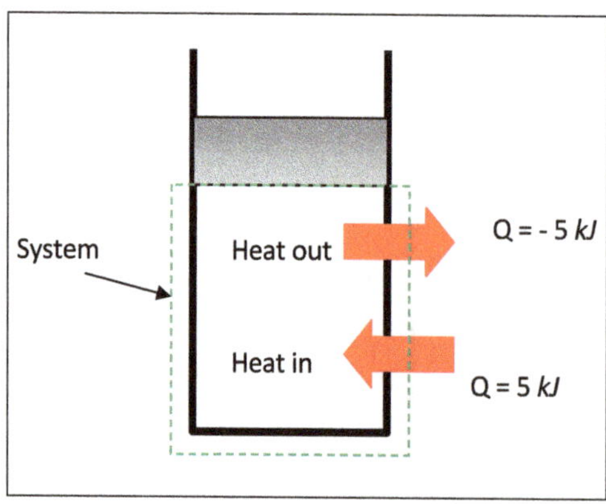

Sign convention for heat and work.

Similarities between work and heat transfer:

- Both are recognized at the boundaries of the system as they cross them (boundary phenomena).
- Systems posses energy, but not heat or work (transfer phenomena).
- Both are associated with a process, not a state. Heat or work has no meaning at a state.
- Both are path functions, their magnitudes depend on the path followed during a process as well as the end states.

Thermodynamic Cycles

In an isolated system, the total energy remains the same. In a thermodynamic cycle, the net heat supplied to the system equals the net work done by the system. Like for example, the batteries we use convert chemical energy into electrical energy. Also the electric bulbs transform electrical energy into light energy.

Work done on, or by a gas, depends not only on the initial and final states of the gas but also on the process, or the path which produces the final state. Similarly, the amount of heat transferred into, or from a gas also depends on the initial and final states and the process which produces the final state.

The internal energy is just a form of energy like the potential energy of an object at some height above the earth, or the kinetic energy of an object in motion. In the same way that potential energy converts in to kinetic energy while conserving the total energy of the system, the internal energy of a thermodynamic system converts in to either kinetic or potential energy. Like potential energy, the internal energy can be stored in the system. The first law of thermodynamics allows for many possible states of a system to exist, but only certain states are found to exist in nature.

Limitations of First Law of Thermodynamics

- The limitation of the first law of thermodynamics is that it does not say anything about the direction of flow of heat.
- It does not say anything whether the process is a spontaneous process or not.
- The reverse process is not possible. In actual practice, the heat doesn't convert completely into work.If it would have been possible to convert the whole heat into work, then we could drive ships across the ocean by extracting heat from the water of the ocean.

Example: Find out the internal energy of a system which has constant volume and the heat around the system is increased by 50 J?

Solution: Given, $q = 50$ J

Since the gas is in constant volume, $\Delta v = 0$

$$w = p\Delta v = 0$$

The equation for internal energy is, $\Delta U = q + w$.

$$\Delta U = q + 0$$

$$\Delta U = q = 50\ J$$

Electrical Work

The work that is done on a system by electrons. When N coulombs of electrons move through a potential difference V, the electrical work done is:

$$W_e = VN(kJ)$$

Which can be explained in the rate form as:

$$\dot{W_e} = VI \quad (kW)$$

Example:

A well-insulated electrical oven is being heated through its heating element. Determine whether it is work or heat interaction. Consider two systems: a) the entire oven (including the heater), and b) only the air in the oven (without the heater).

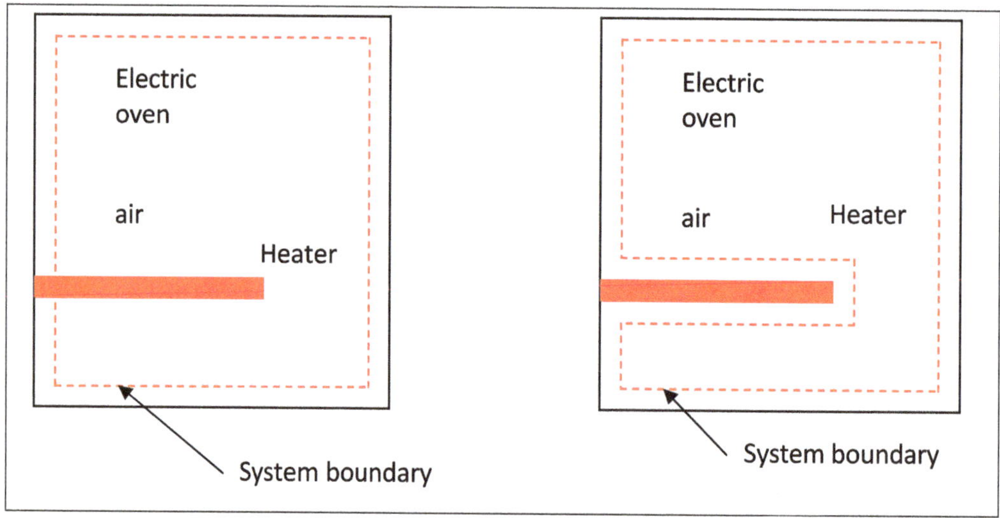

Schematic for example.

Solution:

The energy content of the oven is increased during this process:

- The energy transfer to the oven is not caused by a temperature difference between the oven and air. Instead, it is caused by electrical energy crossing the system boundary and thus: this is a work transfer process.
- This time, the system boundary includes the outer surface of the heater and will not cut through it. Therefore, no electrons will be crossing the system boundary. Instead, the

energy transfer is a result of a temperature difference between the electrical heater and air, thus: this is a heat transfer process.

Mechanical Work

There are several ways of doing work, each in some way related to a force acting through a distance.

$$W = F.s \quad (kJ)$$

If the force is not constant, we need to integrate:

$$W = \int_1^2 F.ds \quad (kJ)$$

There are two requirements for a work interaction:

- There must be a force acting on the boundary.
- The boundary must move.

Therefore, the displacement of the boundary without any force to oppose or drive this motion (such as expansion of a gas into evacuated space) is not a work interaction, W = 0.

Also, if there are no displacements of the boundary, even if an acting force exists, there will be no work transfer W = 0 (such as increasing gas pressure in a rigid tank).

Moving Boundary Work

The expansion and compression work is often called moving boundary work, or simply boundary work.

We analyze the moving boundary work for a quasi-equilibrium process. Consider the gas enclosed in a piston-cylinder at initial P and V. If the piston is allowed to move a distance ds in a quasi-equilibrium manner, the differential work is:

$$\delta W_b = F.ds = PAds = PdV$$

The quasi-equilibrium expansion process is shown in figure. On this diagram, the differential area dA under the process curve in P-V diagram is equal to PdV, which is the differential work.

A gas can follow several different paths from state 1 to 2, and each path will have a different area underneath it (work is path dependent).

The net work or cycle work is shown in figure. In a cycle, the net change for any properties (point functions or exact differentials) is zero. However, the net work and heat transfer depend on the cycle path.

$$\Delta U = \Delta P = \Delta T = \Delta(\text{any property}) = 0 \text{ for a cycle}$$

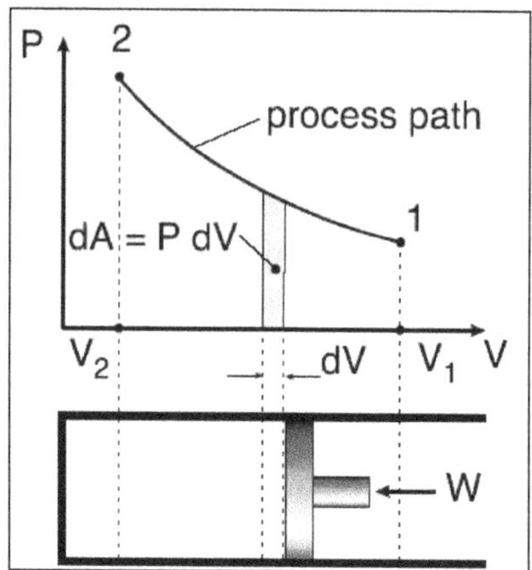

Fig.: the area under P-V diagram represents the boundary work.

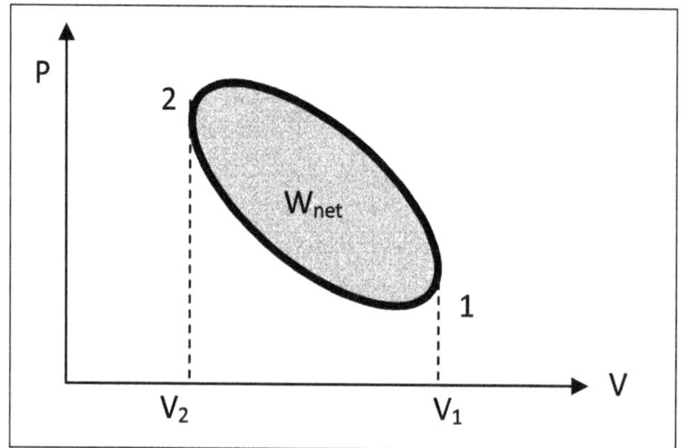

Network done during a cycle.

Polytropic Process

During expansion and compression processes of real gases, pressure and volume are often related by $PV^n = C$, where n and C are constants. The moving work for a polytropic process can be found:

$$W_{polytopic} = \int_1^2 PdV = \int_1^2 CV^{-n}dV = \frac{P_2V_2 - P_1V_1}{1-n}$$

Since $P_1V_1^n = P_2V_2^n = C$. For an ideal gas (PV = mRT) it becomes:

$$W_{polytropic} = \frac{mR(T_2 - T_1)}{1-n}, \quad n \neq 1 \quad (kJ)$$

The special case n = 1 is the isothermal expansion $P_1V_1 = P_2V_2 = mRT_0 = C$, which can be found from:

$$W_{b,isothermal} = \int_1^2 PdV = \int_1^2 \frac{C}{V}dV = P_1V_1 \ \text{In}\left(\frac{V_2}{V_1}\right), \quad n = 1 \ (kJ)$$

Since for an ideal gas, $PV=mRT_0$ at constant temperature T_0, or P = C/V.

Example of Polytropic Work

A gas in piston-cylinder assembly undergoes a polytropic expansion. The initial pressure is 3 bar, the initial volume is 0.1 m³, and the final volume is 0.2 m³. Determine the work for the process, in kJ, if a) n = 1.5, b) n = 1.0, and c) n = 0.

Solution:

Assume that i) the gas is a closed system, ii) the moving boundary is only work mode, and iii) the expansion is polytropic.

a) n = 1.5

$$W = \int_{V1}^{V2} PdV = \frac{P_2V_2 - P_1V_1}{1-n}$$

We need P2 that can be found from $P_1V_1^n = P_2V_2^n$:

$$P_2 = P_1\left(\frac{V_1}{V_2}\right)^n = (3\,bar)\left(\frac{0.1}{0.2}\right)^{1.5} = 1.06\ bar$$

$$W = \left(\frac{(1.06\,bar)(0.2\,m^3) - (3)(0.1)}{1-1.5}\right)\left(\frac{10^5\,N/m^2}{1\ bar}\right)\left(\frac{1\ kJ}{10^3\ N.m}\right) = 17.6\ kJ$$

b) n =1, the pressure volume relationship is PV = constant. The work is:

$$W = \int_1^2 PdV = P_1V_1\,\text{In}\left(\frac{V_2}{V_1}\right)$$

$$W = (3\ bar)(0.1\,m^3)\left[\frac{10^5\,N/3 = m^2}{1\,bar}\right]\left(\frac{1\ kJ}{10^3\ N.m}\right)\text{In}\left(\frac{0.2}{0.1}\right) = 20.79\ kJ$$

c) For n = 0, the pressure-volume relation reduces to P=constant (isobaric process) and the integral become W= P ($V_2 - V_1$).

Substituting values and converting units as above, W = 30 kJ.

Spring Work

For linear elastic springs, the displacement x is proportional to the force applied:

$$F = k_s x$$

where k_s is the spring constant and has the unit kN/m. The displacement x is measured from the undisturbed position of the spring. The spring work is:

$$W_{spring} = \frac{1}{2}k_s\left(x_2^2 - x_1^2\right)\ (kJ)$$

The work done on a spring equals the energy stored in the spring.

Non-mechanical Forms of Work

Non-mechanical forms of work can be treated in a similar manner to mechanical work. Specify a generalized force F acting in the direction of a generalized displacement x, the work transfer associated with the displacement dx is:

$$\delta W = F.dx$$

Example: Mechanical work

Calculate the work transfer in the following process:

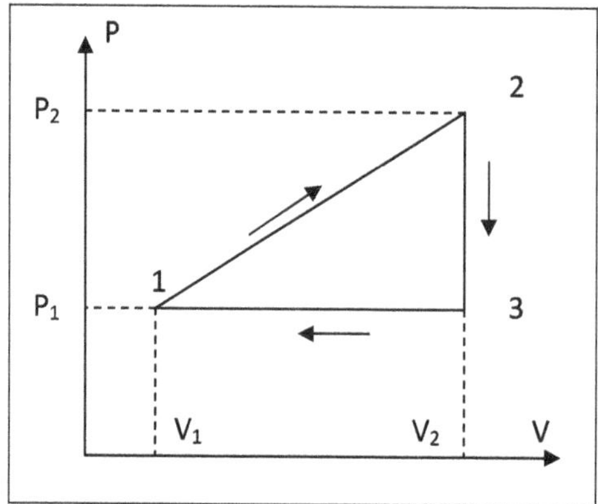

Schematic P-V diagram for Example.

Solution:

Process 1-2 is an expansion ($V_2 > V_1$) and the system is doing work ($W_{12} > 0$), thus:

$$W_{12} = P_1(V_2 - V_1) + [0.5(P_1 + P_2) - P_1](V_2 - V_1)$$
$$= (V_2 - V_1)(P_1 + P_2)/2$$

Process 2 - 3 is an isometric process (constant volume $V_3 = V_2$), so $W_{23} = 0$

Process 3 - 1 is a compression ($V_3 > V_1$), work is done on the system, ($W_{31} < 0$)

$W_{31} = - P_1(V_2 - V_1)$

$W_{cycle} = W_{net} = W_{12} + W_{23} + W_{31} = (V_2 - V_1)(P_2 - P_1)/2$

Note that in a cycle $\Delta U = \Delta P = \Delta T = \Delta(\text{any property}) = 0$

Example of Fist law

Air is contained in a vertical piston-cylinder assembly fitted with an electrical resistor. The atmospheric pressure is 100 kPa and piston has a mass of 50 kg and a face area of 0.1 m². Electric current passes through the resistor, and the volume of air slowly increases by 0.045 m³. The mass of the air is 0.3 kg and its specific energy increases by 42.2 kJ/kg. Assume the assembly (including the piston) is insulated and neglect the friction between the cylinder and piston, g = 9.8 m/s². Determine the heat transfer from the resistor to air for a system consisting a) the air alone, b) the air and the piston.

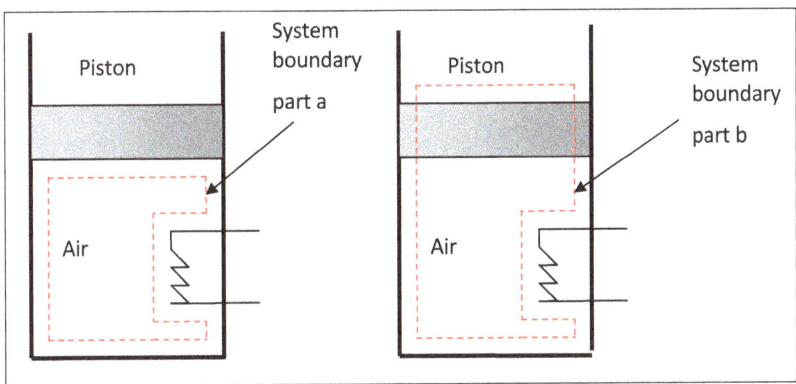

Schematic for problem.

Assumptions:

- Two closed systems are under consideration, as shown in schematic.
- The only heat transfer is from the resistor to the air. ΔKE = ΔPE = 0 (for air).
- The internal energy of the piston is not affected by the heat transfer.

a) Taking the air as the system,

$$(\Delta KE + \Delta PE + \Delta U)_{air} = Q - W$$

$$Q = W + \Delta U_{air}$$

For this system work is done at the bottom of the piston. The work done by the system is (at constant pressure):

$$W = \int_{V1}^{V2} PdV = P(V_2 - V_1)$$

The pressure acting on the air can be found from:

$$PA_{piston} = m_{piston}\, g + P_{atm}\, A_{piston}$$

$$P = \frac{m_{piston}\, g}{A_{piston}} + P_{atm}$$

$$P = \frac{(50\,kg)(9.81\,m/s^2)}{(0.1\,m^2)}\left(\frac{1\,Pa}{1\,N/m^2}\right)\left(\frac{1\,kPa}{1000\,Pa}\right) + 100\ kPa = 104.91\ kPa$$

Thus, the work is

$$W = (104.91\ \text{kPa})(0.045\text{m}^3) = 4.721\ \text{kJ}$$

With $\Delta U_{air} = m_{air}\ \Delta u_{air}$, the heat transfer is

$$Q = W + m_{air}\ \Delta u_{air} = 4.721\ \text{kJ} + (0.3\ \text{kg})(42.2\ \text{kJ/kg}) = 17.38\ \text{kJ}$$

b) System consisting the air and the piston. The first law becomes:

$$(\Delta KE + \Delta PE + \Delta U)_{air} + (\Delta KE + \Delta PE + \Delta U)_{piston} = Q - W$$

where $(\Delta KE = \Delta PE)_{air} = 0$ and $(\Delta KE = \Delta U)_{piston} = 0$. Thus, it simplifies to:

$$(\Delta U)_{air} + (\Delta PE)_{piston} = Q - W$$

For this system, work is done at the top of the piston and pressure is the atmospheric pressure. The work becomes:

$$W = P_{atm}\ \Delta V = (100\ \text{kPa})(0.045\text{m}^3) = 4.5\ \text{kJ}$$

The elevation change required to evaluate the potential energy change of the piston can be found from the volume change:

$$\Delta z = \Delta V / A_{piston} = 0.045\ \text{m}^3 / 0.1\ \text{m}^2 = 0.45\ \text{m}$$

$$(\Delta PE)_{piston} = m_{piston}\ g\Delta z = (50\ \text{kg})(9.81\ \text{m/s}^2)(0.45\ \text{m}) = 220.73\ \text{J} = 0.221\ \text{kJ}$$

$$Q = W + (\Delta PE)_{piston} + m_{air}\ \Delta u_{air}$$

$$Q = 4.5\ \text{kJ} + 0.221\ \text{kJ} + (0.3\ \text{kg})(42.2\ \text{kJ/kg}) = 17.38\ \text{kJ}$$

Note that the heat transfer is identical in both systems.

Specific Heat

The specific heat is defined as the energy required to raise the temperature of a unit mass of a substance by one degree. There are two kinds of specific heats:

- Specific heat at constant volume, C_v (the energy required when the volume is maintained constant)
- Specific heat at constant pressure, C_p (the energy required when the pressure is maintained constant)

The specific heat at constant pressure C_p is always higher than C_v because at constant pressure the system is allowed to expand and energy for this expansion must also be supplied to the system.

Let's consider a stationary closed system undergoing a constant-volume process ($w_b = 0$). Applying the first law in the differential form:

$$\delta q - \delta w = du$$

At constant volume (no work) and by using the definition of C_v, one can write:

$$C_v dT = du$$

or

$$C_v = \left(\frac{\partial u}{\partial T}\right)_v$$

Similarly, an expression for the specific heat at constant pressure Cp can be found. From the first law, for a constant pressure process ($w_b + \Delta u = \Delta h$). It yields:

$$C_p = \left(\frac{\partial h}{\partial T}\right)_p$$

- Specific heats (both C_v and C_p) are properties and therefore independent of the type of processes.
- C_v is related to the changes in internal energy u, and C_p to the changes in enthalpy, h.

It would be more appropriate to define: C_v is the change in specific internal energy per unit change in temperature at constant volume. C_p is the change in specific enthalpy per unit change in temperature at constant pressure.

Specific Heats for Ideal Gases

It has been shown mathematically and experimentally that the internal energy is a function of temperature only.

$$u = u(T)$$

Using the definition of enthalpy ($h = u + Pv$) and the ideal gas equation of state ($Pv = RT$), we have:

$$h = u + RT$$

Since R is a constant and u is a function of T only:

$$h = h(T)$$

Therefore, at a given temperature, u, h, C_v and C_p of an ideal gas will have fixed values regardless of the specific volume or pressure. For an ideal gas, we have:

$$du = C_v(T)\,dT$$
$$dh = C_p(T)\,dT$$

The changes in internal energy or enthalpy for an ideal gas during a process are determined by integrating:

$$\Delta u = u_2 - u_1 = \int_1^2 C_v(T)\,dT \quad (kJ/kg)$$

$$\Delta h = h_2 - h_1 = \int_1^2 C_p(T)\,dT \quad (kJ/kg)$$

As low pressures, all real gases approach ideal-gas behavior, and therefore their specific heats depend on temperature only. The specific heats of real gases at low pressures are called ideal-gas specific heats (or zero-pressure specific heats) and are often denoted by C_{po} and C_{vo}. To carry out the above integrations, we need to know $C_v(T)$ and $C_p(T)$.

For an ideal gas, we can write:

$$RT = h(T) - u(T)$$

$$R = \frac{dh}{dT} - \frac{du}{dT}$$

$$R = C_p - C_v$$

The ratio of specific heats is called the specific heat ratio $k = C_p/C_v$:

- Varies with temperature, but this variation is very mild.
- For monatomic gases, its value is essentially constant at 1.67.
- Many diatomic gases, including air, have a specific heat ratio of about 1.4 at room temperature.

Specific Heats for Solids and Liquids

A substance whose specific volume (or density) is constant is called incompressible substance. The specific volumes of solids and liquids (which can be assumed as incompressible substances) essentially remain constant during a process.

The constant volume assumption means that the volume work (boundary work) is negligible compared with other forms of energy. As a result, it can be shown that the constant-volume and constant-pressure specific heats are identical for incompressible substances:

$$C_p = C_v = C$$

Specific heats of incompressible substances are only a function of temperature,

$$C = C(T)$$

The change of internal energy between state 1 and 2 can be obtained by integration:

$$\Delta u = u_2 - u_1 = \int_1^2 C(T)\,dT \quad (kJ/kg)$$

For small temperature intervals, a C at averaged temperature can be used and treated as a constant, yielding:

$$\Delta u \approx C_{ave}\left(T_2 - T_1\right)$$

The enthalpy change of incompressible substance can be determined from the definition of enthalpy (h = u + Pv)

$$h_2 - h_1 = \left(u_2 - u_1\right) + v\left(P_2 - P_1\right)$$
$$\Delta h = \Delta u + v\Delta P \quad \left(kJ/kg\right)$$

The term vΔP is often small and can be neglected, so Δh = Δu.

Example of Specific Heat and First Law

Two tanks are connected by a valve. One tank contains 2 kg of CO_2 at 77 °C and 0.7 bar. The other tank has 8 kg of the same gas at 27 °C and 1.2 bar. The valve is opened and gases are allowed to mix while receiving energy by heat transfer from the surroundings. The final equilibrium temperature is 42°C. Using ideal gas model, determine a) the final equilibrium pressure b) the heat transfer for the process.

Assumptions:

- The total amount of CO_2 remains constant (closed system).
- Ideal gas with constant C_v.
- The initial and final states in the tanks are equilibrium. No work transfer.

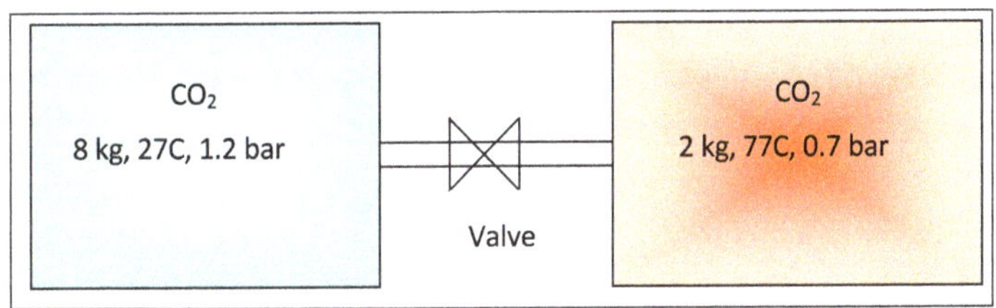

The final pressure can be found from ideal gas equation of state:

$$P_f = \frac{m_t R T_f}{V_1 + V_2} = \frac{\left(m_1 + m_2\right) R T_f}{V_1 + V_2}$$

For tank 1 and 2, we can write: V1 = m_1RT_1/P_1 and $V_2 = m_2RT_2/P_2$. Thus, the final pressure, P_f becomes:

$$P_f = \frac{\left(m_1 + m_2\right) R T_f}{\left(\frac{m_1 R T_1}{P_1}\right) + \left(\frac{m_2 R T_2}{P_2}\right)} = \frac{\left(m_1 + m_2\right) T_f}{\left(\frac{m_1 T_1}{P_1}\right) + \left(\frac{m_2 T_2}{P_2}\right)}$$

$$P_f = \frac{\left(100\,kg\right)\left(315\,K\right)}{\frac{\left(2\,kg\right)\left(350\,K\right)}{0.7\ bar} + \frac{\left(8\,kg\right)\left(300\,K\right)}{1.2\,bar}} = 1.05\ bar$$

b) The heat transfer can be found from an energy balance:

$$\Delta U = Q - W$$

With W = 0,

$$Q = U_f - U_i$$

where initial internal energy is: $U_i = m_1 u(T_1) + m_2 u(T_2)$

The final internal energy is: $U_f = (m_1 + m_2) u(T_f)$

The energy balance becomes:

$$Q = m_1 [u(T_f) - u(T_1)] + m_2 [u(T_f) - u(T_2)]$$

Since the specific heat C_v is constant

$$Q = m_1 C_v [T_f - T_1] + m_2 C_v [T_f - T_2]$$

$$Q = (2\,kg)\left(0.745\frac{kJ}{kg.K}\right)(315\,K - 350\,K) + (8kg)\left(0.745\frac{kJ}{kg.K}\right)(315\,K - 300\,K) = 37.25\,kJ$$

The plus sign indicates that the heat transfer is into the system.

Human Metabolism and the First Law of Thermodynamics

Human metabolism is the conversion of food into heat transfer, work, and stored fat. Metabolism is an interesting example of the first law of thermodynamics in action. We now take another look at these topics via the first law of thermodynamics. Considering the body as the system of interest, we can use the first law to examine heat transfer, doing work, and internal energy in activities ranging from sleep to heavy exercise. What are some of the major characteristics of heat transfer, doing work, and energy in the body? For one, body temperature is normally kept constant by heat transfer to the surroundings. This means Q is negative. Another fact is that the body usually does work on the outside world. This means W is positive. In such situations, then, the body loses internal energy, since $\Delta U = Q - W$ is negative.

Now consider the effects of eating. Eating increases the internal energy of the body by adding chemical potential energy (this is an unromantic view of a good steak). The body metabolizes all the food we consume. Basically, metabolism is an oxidation process in which the chemical potential energy of food is released. This implies that food input is in the form of work. Food energy is reported in a special unit, known as the Calorie. This energy is measured by burning food in a calorimeter, which is how the units are determined.

In chemistry and biochemistry, one calorie (spelled with a lowercase c) is defined as the energy (or heat transfer) required to raise the temperature of one gram of pure water by one degree Celsius. Nutritionists and weight-watchers tend to use the dietary calorie, which is frequently called a Calorie (spelled with a capital C). One food Calorie is the energy needed to raise the temperature of one kilogram of water by one degree Celsius. This means that one dietary Calorie is equal to one kilocalorie for the chemist, and one must be careful to avoid confusion between the two.

Again, consider the internal energy the body has lost. There are three places this internal energy can go—to heat transfer, to doing work, and to stored fat (a tiny fraction also goes to cell repair and growth). Heat transfer and doing work take internal energy out of the body, and food puts it back. If you eat just the right amount of food, then your average internal energy remains constant. Whatever you lose to heat transfer and doing work is replaced by food, so that, in the long run, $\Delta U = 0$. If you overeat repeatedly, then ΔU is always positive, and your body stores this extra internal energy as fat. The reverse is true if you eat too little. If ΔU is negative for a few days, then the body metabolizes its own fat to maintain body temperature and do work that takes energy from the body. This process is how dieting produces weight loss.

Life is not always this simple, as any dieter knows. The body stores fat or metabolizes it only if energy intake changes for a period of several days. Once you have been on a major diet, the next one is less successful because your body alters the way it responds to low energy intake. Your basal metabolic rate (BMR) is the rate at which food is converted into heat transfer and work done while the body is at complete rest. The body adjusts its basal metabolic rate to partially compensate for over-eating or under-eating. The body will decrease the metabolic rate rather than eliminate its own fat to replace lost food intake. You will chill more easily and feel less energetic as a result of the lower metabolic rate, and you will not lose weight as fast as before. Exercise helps to lose weight, because it produces both heat transfer from your body and work, and raises your metabolic rate even when you are at rest. Weight loss is also aided by the quite low efficiency of the body in converting internal energy to work, so that the loss of internal energy resulting from doing work is much greater than the work done. It should be noted, however, that living systems are not in thermal equilibrium.

The body provides us with an excellent indication that many thermodynamic processes are irreversible. An irreversible process can go in one direction but not the reverse, under a given set of conditions. For example, although body fat can be converted to do work and produce heat transfer, work done on the body and heat transfer into it cannot be converted to body fat. Otherwise, we could skip lunch by sunning ourselves or by walking down stairs. Another example of an irreversible thermodynamic process is photosynthesis. This process is the intake of one form of energy—light—by plants and its conversion to chemical potential energy. One great advantage of conservation laws such as the first law of thermodynamics is that they accurately describe the beginning and ending points of complex processes, such as metabolism and photosynthesis, without regard to the complications in between.

(a) The first law of thermodynamics applied to metabolism. Heat transferred out of the body (Q) and work done by the body (W) remove internal energy, while food intake replaces it. (Food intake may be considered as work done on the body.) (b) Plants convert part of the radiant heat transfer in sunlight to stored chemical energy, a process called photosynthesis.

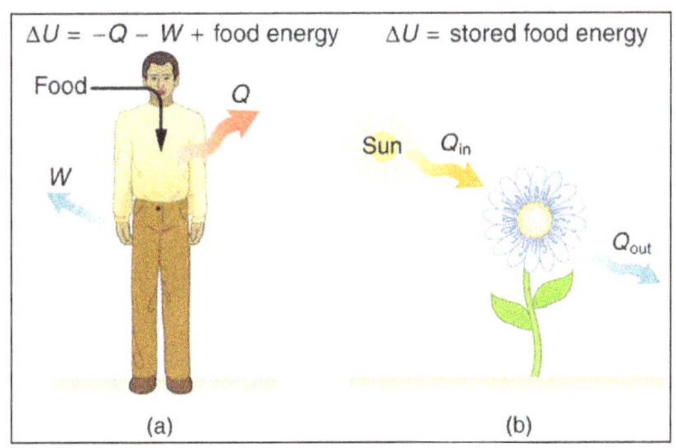

Second Law of Thermodynamics

The second law of thermodynamics states that there is never a decrease in the entropy of an isolated system. In other words, the entropy change of an isolated system can never hold a negative value. However, the entropy can remain constant if the system is in a state of thermodynamic equilibrium.

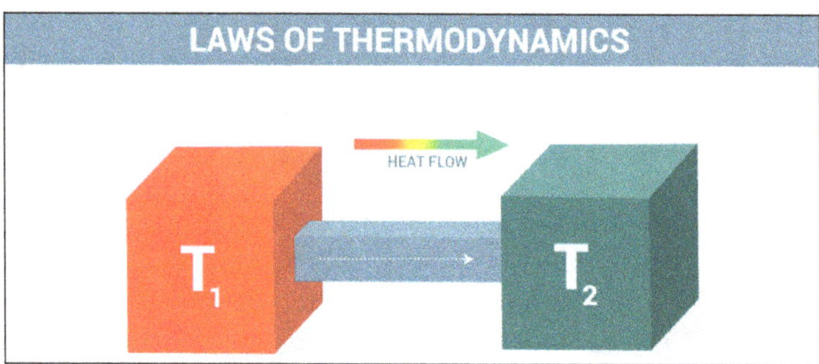

In an ideal reversible process, the overall entropy remains the same. An increase in the entropy results in an irreversible change, resulting in an asymmetry between the previous and the future states of the body.

Statements of the Second Law of Thermodynamics

Many different statements of the second law of thermodynamics have been formulated. One of the earliest statements of this law was formulated by Nicolas Leonard Sadi Carnot, a French scientist who is often referred to as the 'father of thermodynamics'.

Carnot's Principle and Theorem

- Carnot's principle states that the efficiency of a cycle of a Carnot engine is solely dependent on the temperatures of the heat reservoirs.
- It also states that this Carnot engine has the highest possible efficiency a heat engine can have using these two temperatures.
- Carnot's Theorem states that the efficiency of all irreversible heat engines that operate between two heat reservoirs is less than that of a Carnot engine.
- It goes on to state that the efficiency of a reversible heat engine that operates between two heat reservoirs is equal to that of a Carnot engine.

Clausius Statement

- The relationship between heat and work was examined by the German physicist Rudolf Clausius in his statement of the second law of thermodynamics.
- He stated that heat can never pass from a cold body to a hotter body unless some work is done by an external source of energy.
- For example, in a refrigerator, the heat flows from a cold region to a hot region due to the work performed by the refrigerating system.

Kelvin-Planck Statement

- This statement was derived from the independent statements of the second law put forth by Lord Kelvin and Max Planck.
- It states that it is impossible to construct a device that operates via a thermodynamic cycle and converts all the thermal energy from a heat reservoir into work.
- This statement of the second law of thermodynamics dictates that all of the heat absorbed from heat source cannot be converted into work and that some of the heat must be passed on to a colder body.

Another important principle related to the second law is Planck's principle, which states that an adiabatic process (a process that occurs without a transfer of heat or mass between the system and its surroundings) causes an increase in the internal energy of a closed system.

Derivation of the Second Law of Thermodynamics

Entropy is the degree of randomness or disorder in a system. In thermodynamics, this quantity is a measure of the thermal energy in a system that cannot be used to perform work. The entropy of a system can be expressed in terms of the changes undergone by the system when its initial and final states are compared.

In order to measure this quantity, it is expressed in terms of entropy change (denoted by ΔS). Some important points detailing the relationship between entropy and the second law of thermodynamics are listed below:

- An isentropic process is a process in which the overall entropy of the system remains constant, i.e. ΔS = 0. This implies the entropy of the initial and final states of the system holds a constant value.
- The mass of a closed system generally remains constant. However, an exchange of heat between the system and the surrounding can still occur. This implies that a change in the entropy of a system arises when the heat content of a closed system is disturbed by its surroundings.
- The movement of molecules in a closed system leads to a transfer of energy. This can, in turn, lead to an increase in the entropy of the system.

The Clausius concept of entropy is used to measure the direction of a spontaneous change, stating that the spontaneous changes caused by an irreversible process must proceed in the direction of increasing entropy.

The change in the entropy of the universe is given by the sum of the entropy change of the system and its surroundings. This can be equated as follows:

$$\Delta S_{universe} = \Delta S_{system} + \Delta S_{surrounding} = \frac{q_{surrounding}}{T} + \frac{q_{surrounding}}{T}$$

where ΔS is the entropy change, q is the heat absorbed, and T is the temperature. The equation for the heat energy absorbed by the system in a reversible and isothermal (no change in temperature) process is given by:

$$q = nRT\ In\frac{V_2}{V_1}$$

In an isothermal process, the total amount of heat absorbed by the system must be equal to the total amount of heat lost by the surrounding. This implies that $q_{system} = -q_{surrounding}$.

Therefore, the change in the entropy of the universe can be expressed as:

$$\Delta S_{universe} = \frac{nRT\ In\frac{V_2}{V_1}}{T} + \frac{-nRT\ In\frac{V_2}{V_1}}{T} = 0$$

This implies that the entropy change in a reversible process is zero. Now, if the process is irreversible, the entropy change must be greater than that of a reversible process, i.e.

$$\Delta S_{universe} = \frac{nRT\ In\frac{V_2}{V_1}}{T} > 0$$

Combining the equation for entropy for a reversible and irreversible process, the following equation can be obtained:

$$\Delta S_{universe} = \Delta S_{system} + \Delta S_{surroundings} \geq 0$$

Which is the equation for the second law of thermodynamics, i.e. the change in the entropy of an isolated system cannot be negative.

Statistical Mechanics

Statistical mechanics gives an explanation for the second law by postulating that a material is composed of atoms and molecules which are in constant motion. A particular set of positions and velocities for each particle in the system is called a microstate of the system and because of the constant motion, the system is constantly changing its microstate. Statistical mechanics postulates that, in equilibrium, each microstate that the system might be in is equally likely to occur, and when this assumption is made, it leads directly to the conclusion that the second law must hold in a statistical sense. That is, the second law will hold on average, with a statistical variation on the order of $1/\sqrt{N}$ where *N* is the number of particles in the system. For everyday (macroscopic) situations, the probability that the second law will be violated is practically zero. However, for systems with a small number of particles, thermodynamic parameters, including the entropy, may show significant statistical deviations from that predicted by the second law. Classical thermodynamic theory does not deal with these statistical variations.

Derivation from Statistical Mechanics

The first mechanical argument of the Kinetic theory of gases that molecular collisions entail an equalization of temperatures and hence a tendency towards equilibrium was due to James Clerk Maxwell in 1860; Ludwig Boltzmann with his H-theorem of 1872 also argued that due to collisions gases should over time tend toward the Maxwell-Boltzmann distribution.

Due to Loschmidt's paradox, derivations of the Second Law have to make an assumption regarding the past, namely that the system is uncorrelated at some time in the past; this allows for simple probabilistic treatment. This assumption is usually thought as a boundary condition, and thus the second Law is ultimately a consequence of the initial conditions somewhere in the past, probably at the beginning of the universe (the Big Bang), though other scenarios have also been suggested.

Given these assumptions, in statistical mechanics, the Second Law is not a postulate, rather it is a consequence of the fundamental postulate, also known as the equal prior probability postulate, so long as one is clear that simple probability arguments are applied only to the future, while for the past there are auxiliary sources of information which tell us that it was low entropy. The first part of the second law, which states that the entropy of a thermally isolated system can only increase, is a trivial consequence of the equal prior probability postulate, if we restrict the notion of the entropy to systems in thermal equilibrium. The entropy of an isolated system in thermal equilibrium containing an amount of energy of E is:

$$S = k_B \ln\left[\Omega(E)\right]$$

Where $\Omega(E)$ is the number of quantum states in a small interval between E and $E+\delta E$. Here δE is a macroscopically small energy interval that is kept fixed. Strictly speaking this means that the entropy depends on the choice of δE. However, in the thermodynamic limit (i.e. in the limit of infinitely large system size), the specific entropy (entropy per unit volume or per unit mass) does not depend on δE.

Suppose we have an isolated system whose macroscopic state is specified by a number of variables. These macroscopic variables can, e.g., refer to the total volume, the positions of pistons in the system, etc. Then Ω will depend on the values of these variables. If a variable is not fixed, (e.g. we do not clamp a piston in a certain position), then because all the accessible states are equally likely in equilibrium, the free variable in equilibrium will be such that Ω is maximized as that is the most probable situation in equilibrium.

If the variable was initially fixed to some value then upon release and when the new equilibrium has been reached, the fact the variable will adjust itself so that Ω is maximized, implies that the entropy will have increased or it will have stayed the same (if the value at which the variable was fixed happened to be the equilibrium value). Suppose we start from an equilibrium situation and we suddenly remove a constraint on a variable. Then right after we do this, there are a number Ω of accessible microstates, but equilibrium has not yet been reached, so the actual probabilities of the system being in some accessible state are not yet equal to the prior probability of $1/\Omega$. We have already seen that in the final equilibrium state, the entropy will have increased or have stayed the same relative to the previous equilibrium state. Boltzmann's H-theorem, however, proves that the quantity H increases monotonically as a function of time during the intermediate out of equilibrium state.

Derivation of the Entropy Change for Reversible Processes

The second part of the Second Law states that the entropy change of a system undergoing a reversible process is given by:

$$dS = \frac{\delta Q}{T}$$

Where the temperature is defined as:

$$\frac{1}{k_B T} \equiv \beta \equiv \frac{d \ln\left[\Omega(E)\right]}{dE}$$

Suppose that the system has some external parameter, x, that can be changed. In general, the energy eigenstates of the system will depend on x. According to the adiabatic theorem of quantum mechanics, in the limit of an infinitely slow change of the system's Hamiltonian, the system will stay in the same energy eigenstate and thus change its energy according to the change in energy of the energy eigenstate it is in.

The generalized force, X, corresponding to the external variable x is defined such that Xdx is the work performed by the system if x is increased by an amount dx. E.g., if x is the volume, then X is the pressure. The generalized force for a system known to be in energy eigenstate E_r is given by:

$$X = -\frac{dE_r}{dx}$$

Since the system can be in any energy eigenstate within an interval of δE, we define the generalized force for the system as the expectation value of the above expression:

$$X = -\left\langle \frac{dE_r}{dx} \right\rangle$$

To evaluate the average, we partition the $\Omega(E)$ energy eigenstates by counting how many of them have a value for $\frac{dE_r}{dx}$ within a range between Y and $Y + \delta Y$. Calling this number $\Omega_Y(E)$, we have:

$$\Omega(E) = \sum_Y \Omega_Y(E)$$

The average defining the generalized force can now be written:

$$X = -\frac{1}{\Omega(E)} \sum_Y Y \Omega_Y(E)$$

We can relate this to the derivative of the entropy with respect to x at constant energy E as follows. Suppose we change x to x + dx. Then $\Omega(E)$ will change because the energy eigenstates depend on x, causing energy eigenstates to move into or out of the range between E and $E + \delta E$. Let's focus again on the energy eigenstates for which $\frac{dE_r}{dx}$ lies within the range between Y and $Y + \delta Y$. Since these energy eigenstates increase in energy by Y dx, all such energy eigenstates that are in the interval ranging from E – Y dx to E move from below E to above E. There are:

$$N_Y(E) = \frac{\Omega_Y(E)}{\delta E} Ydx$$

Such energy eigenstates. If $Ydx \leq \delta E$, all these energy eigenstates will move into the range between E and $E + \delta E$ and contribute to an increase in Ω. The number of energy eigenstates that move from below $E + \delta E$ to above $E + \delta E$ is given by $N_Y(E + \delta E)$. The difference:

$$N_Y(E) - N_Y(E + \delta E)$$

is thus the net contribution to the increase in Ω. Note that if Y dx is larger than δE there will be the energy eigenstates that move from below E to above $E + \delta E$. They are counted in both $N_Y(E)$ and $N_Y(E + \delta E)$, therefore the above expression is also valid in that case.

Expressing the above expression as a derivative with respect to E and summing over Y yields the expression:

$$\left(\frac{\partial \Omega}{\partial x}\right)_E = -\sum_Y Y \left(\frac{\partial \Omega_Y}{\partial E}\right)_x = \left(\frac{\partial (\Omega X)}{\partial E}\right)_x$$

The logarithmic derivative of Ω with respect to x is thus given by:

$$\left(\frac{\partial \ln(\Omega)}{\partial x}\right)_E = \beta X + \left(\frac{\partial X}{\partial E}\right)_x$$

The first term is intensive, i.e. it does not scale with system size. In contrast, the last term scales as the inverse system size and will thus vanishes in the thermodynamic limit. We have thus found that:

$$\left(\frac{\partial S}{\partial x}\right)_E = \frac{X}{T}$$

Combining this with

$$\left(\frac{\partial S}{\partial E}\right)_x = \frac{1}{T}$$

Gives:

$$dS = \left(\frac{\partial S}{\partial E}\right)_x dE + \left(\frac{\partial S}{\partial x}\right)_E dx = \frac{dE}{T} + \frac{X}{T} dx = \frac{\delta Q}{T}$$

Derivation for Systems Described by the Canonical Ensemble

If a system is in thermal contact with a heat bath at some temperature T then, in equilibrium, the probability distribution over the energy eigenvalues are given by the canonical ensemble:

$$P_j = \frac{\exp\left(-\frac{E_j}{k_B T}\right)}{Z}$$

Here Z is a factor that normalizes the sum of all the probabilities to 1, this function is known as the partition function. We now consider an infinitesimal reversible change in the temperature and in the external parameters on which the energy levels depend. It follows from the general formula for the entropy:

$$S = -k_B \sum_j P_j \ln\left(P_j\right)$$

that,

$$dS = -k_B \sum_j \ln\left(P_j\right) dP_j$$

Inserting the formula for P_j for the canonical ensemble in here gives:

$$dS = \frac{1}{T}\sum_j E_j dP_j = \frac{1}{T}\sum_j d\left(E_j P_j\right) - \frac{1}{T}\sum_j P_j dE_j = \frac{dE + \delta W}{T} = \frac{\delta Q}{T}$$

Non-equilibrium States

The theory of classical or equilibrium thermodynamics is idealized. A main postulate or assumption, often not even explicitly stated, is the existence of systems in their own internal states of thermodynamic equilibrium. In general, a region of space containing a physical system at a given time, that may be found in nature, is not in thermodynamic equilibrium, read in the most stringent terms. In looser terms, nothing in the entire universe is or has ever been truly in exact thermodynamic equilibrium.

For purposes of physical analysis, it is often enough convenient to make an assumption of thermodynamic equilibrium. Such an assumption may rely on trial and error for its justification. If the assumption is justified, it can often be very valuable and useful because it makes available the theory of thermodynamics. Elements of the equilibrium assumption are that a system is observed to be unchanging over an indefinitely long time, and that there are so many particles in a system, that its particulate nature can be entirely ignored. Under such an equilibrium assumption, in general, there are no macroscopically detectable fluctuations. There is an exception, the case of critical states, which exhibit to the naked eye the phenomenon of critical opalescence. For laboratory studies of critical states, exceptionally long observation times are needed.

In all cases, the assumption of thermodynamic equilibrium, once made, implies as a consequence that no putative candidate "fluctuation" alters the entropy of the system.

It can easily happen that a physical system exhibits internal macroscopic changes that are fast enough to invalidate the assumption of the constancy of the entropy. Or that a physical system has so few particles that the particulate nature is manifest in observable fluctuations. Then the assumption of thermodynamic equilibrium is to be abandoned. There is no unqualified general definition of entropy for non-equilibrium states.

There are intermediate cases, in which the assumption of local thermodynamic equilibrium is a very good approximation, but strictly speaking it is still an approximation, not theoretically ideal.

For non-equilibrium situations in general, it may be useful to consider statistical mechanical definitions of other quantities that may be conveniently called 'entropy', but they should not be confused or conflated with thermodynamic entropy properly defined for the second law. These other quantities indeed belong to statistical mechanics, not to thermodynamics, the primary realm of the second law.

Arrow of Time

Thermodynamic Asymmetry in Time

The second law of thermodynamics is a physical law that is not symmetric to reversal of the time direction. This does not conflict with notions that have been observed of the fundamental laws of physics, namely CPT symmetry, since the second law applies statistically, it is hypothesized, on time-asymmetric *boundary conditions.*

The second law has been proposed to supply a partial explanation of the difference between moving forward and backwards in time, such as why the cause precedes the effect (the causal arrow of time).

Irreversibility

Irreversibility in thermodynamic processes is a consequence of the asymmetric character of thermodynamic operations, and not of any internally irreversible microscopic properties of the bodies. Thermodynamic operations are macroscopic external interventions imposed on the participating bodies, not derived from their internal properties. There are reputed "paradoxes" that arise from failure to recognize this.

Loschmidt's Paradox

Loschmidt's paradox, also known as the reversibility paradox, is the objection that it should not be possible to deduce an irreversible process from the time-symmetric dynamics that describe the microscopic evolution of a macroscopic system.

In the opinion of Schrödinger, "It is now quite obvious in what manner you have to reformulate the law of entropy – or for that matter, all other irreversible statements – so that they be capable of being derived from reversible models. You must not speak of one isolated system but at least of two, which you may for the moment consider isolated from the rest of the world, but not always from each other." The two systems are isolated from each other by the wall, until it is removed by the thermodynamic operation, as envisaged by the law. The thermodynamic operation is externally imposed, not subject to the reversible microscopic dynamical laws that govern the constituents of the systems. It is the cause of the irreversibility. The statement of the law complies with Schrödinger's advice. The cause–effect relation is logically prior to the second law, not derived from it.

Poincaré Recurrence Theorem

The Poincaré recurrence theorem considers a theoretical microscopic description of an isolated physical system. This may be considered as a model of a thermodynamic system after a thermodynamic operation has removed an internal wall. The system will, after a sufficiently long time, return to a microscopically defined state very close to the initial one. The Poincaré recurrence time is the length of time elapsed until the return. It is exceedingly long, likely longer than the life of the universe, and depends sensitively on the geometry of the wall that was removed by the thermodynamic operation. The recurrence theorem may be perceived as apparently contradicting the second law of thermodynamics. More obviously, however, it is simply a microscopic model of thermodynamic equilibrium in an isolated system formed by removal of a wall between two systems. For a typical thermodynamical system, the recurrence time is so large (many many times longer than the lifetime of the universe) that, for all practical purposes, one cannot observe the recurrence. One might wish, nevertheless, to imagine that one could wait for the Poincaré recurrence, and then re-insert the wall that was removed by the thermodynamic operation. It is then evident that the appearance of irreversibility is due to the utter unpredictability of the Poincaré recurrence given only that the initial state was one of thermodynamic equilibrium, as is the case in macroscopic thermodynamics. Even if one could wait for it, one has no practical possibility of picking the right instant at which to re-insert the wall. The Poincaré recurrence theorem provides a solution to Loschmidt's paradox. If an isolated thermodynamic system could be monitored over increasingly many multiples of the average Poincaré recurrence time, the thermodynamic behavior of the system would become invariant under time reversal.

James Clerk Maxwell.

Maxwell's Demon

James Clerk Maxwell imagined one container divided into two parts, *A* and *B*. Both parts are filled with the same gas at equal temperatures and placed next to each other, separated by a wall.

Observing the molecules on both sides, an imaginary demon guards a microscopic trapdoor in the wall. When a faster-than-average molecule from *A* flies towards the trapdoor, the demon opens it, and the molecule will fly from *A* to *B*. The average speed of the molecules in *B* will have increased while in *A* they will have slowed down on average. Since average molecular speed corresponds to temperature, the temperature decreases in *A* and increases in *B*, contrary to the second law of thermodynamics.

One response to this question was suggested in 1929 by Leó Szilárd and later by Léon Brillouin. Szilárd pointed out that a real-life Maxwell's demon would need to have some means of measuring molecular speed, and that the act of acquiring information would require an expenditure of energy.

Maxwell's 'demon' repeatedly alters the permeability of the wall between *A* and *B*. It is therefore performing thermodynamic operations on a microscopic scale, not just observing ordinary spontaneous or natural macroscopic thermodynamic processes.

Reversible Process

A thermodynamic process is reversible if the process can return back in such a that both the system and the surroundings return to their original states, with no other change anywhere else in the universe. It means both system and surroundings are returned to their initial states at the end of the reverse process.

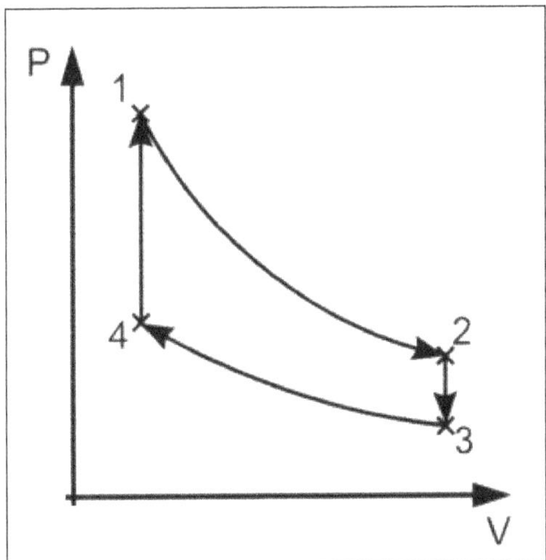

In the figure above, the system has undergone a change from state 1 to state 2. The reversible process can reverse completely and there is no trace left to show that the system had undergone thermodynamic change. During the reversible process, all the changes in state that occur in the system are in thermodynamic equilibrium with each other.

Internally Reversible Process

The process is internally reversible if no irreversibilities occur within the boundaries of the system. In these processes, a system undergoes through a series of equilibrium states, and when the process reverses, the system passes through exactly the same equilibrium states while returning to its initial state.

Externally Reversible Process

In externally reversible process no irreversibilities occur outside the system boundaries during the process. Heat transfer between a reservoir and a system is an externally reversible process if the surface of contact between the system and reservoir is at the same temperature.

A process can be reversible only when its satisfying two conditions:

- Dissipative force must be absent.
- The process should occur in infinite small time.

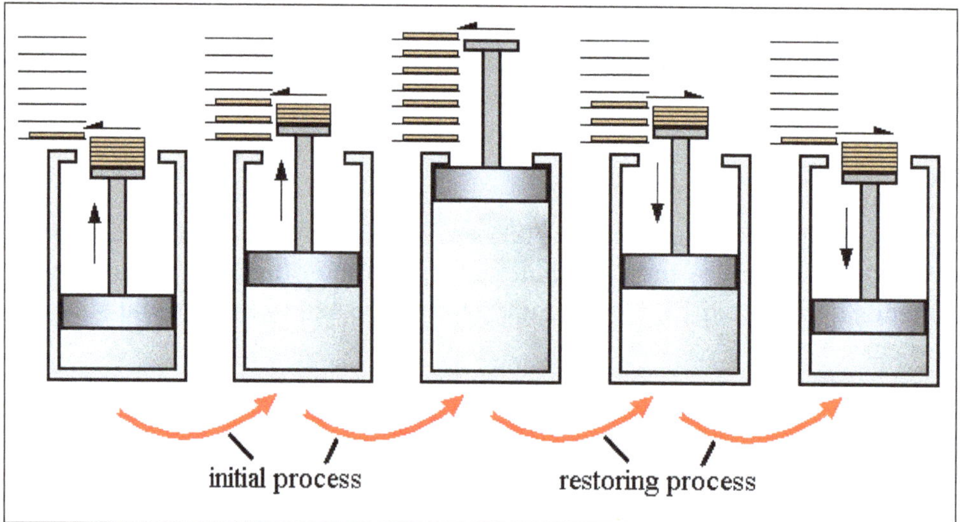

In simple words, the process which can reverse back completely is a reversible process. This means that the final properties of the system can perfectly reverse back to the original properties. The process can be perfectly reversible only if the changes in the process are infinitesimally small. In practical situations it is not possible to trace these extremely small changes in extremely small time, hence the reversible process is also an ideal process. The changes that occur during the reversible process are in equilibrium with each other.

Irreversible Process

An irreversible process is a spontaneous process whose reverse is neither spontaneous nor reversible. That is, the reverse of an irreversible process can never actually occur and is impossible. If a movie is made of a spontaneous process, and the time sequence of the events depicted by the film when it is run backward could not occur in reality, the spontaneous process is irreversible.

Irreversible processes are a result of straying away from the curve, therefore decreasing the amount of overall work done. An irreversible process is a thermodynamic process that departs from equilibrium. In terms of pressure and volume, it occurs when the pressure (or the volume) of a system changes dramatically and instantaneously that the volume (or the pressure) do not have the time to reach equilibrium.

A classic example of an irreversible process is allowing a certain volume of gas to release into a vacuum. By releasing pressure on a sample and allowing it to occupy a large space, the system and surroundings are not in equilibrium during the expansion process.

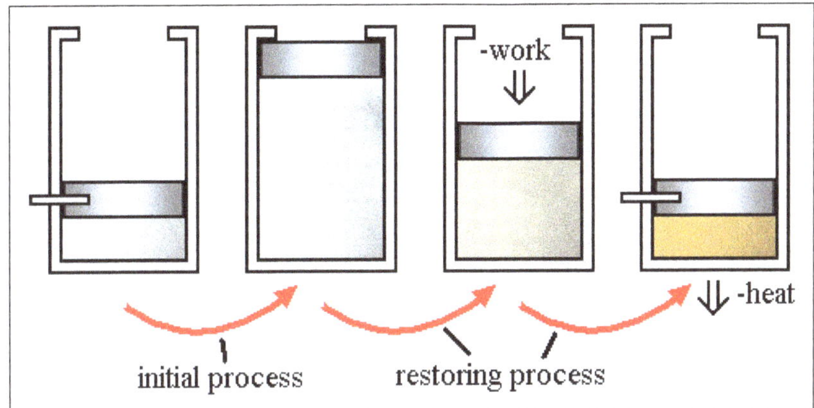

Here little work occurs. However, there is a requirement of significant work, with a corresponding amount of energy dissipation as heat flows to the environment. This is in order to reverse the process.

Heat Engines

A heat engine is a device that converts heat to work. It takes heat from a reservoir then does some work like moving a piston, lifting weight etc. and finally discharging some of the heat energy into the sink.

Heat engines employ a range of methods to apply the heat and to convert the pressure and volume changes into mechanical motion.

From the Gas Laws PV = kNT.

where P is the pressure, V the volume and T the temperature of the gas.

And k is Boltzmann's constant and N is the number of molecules in the gas charge.

Putting energy in the form of heat into a gas will increase its temperature, but at the same time the gas laws mean that the gas pressure or volume or both must increase in proportion. The gas can be restored to its original state by taking this energy out again but not necessarily in the form of heat. The pressure and volume change can be used to perform work by moving a suitably designed mechanical device such as a piston or a turbine blade.

The greater the temperature change, the more energy which can be extracted from the fluid.

Heat Engine as Part of a System

Heat engines enable heat energy to be converted to kinetic energy through the medium of a working fluid.

The diagram shows the system heat flow. Heat is transferred from the source, through working fluid in the heat engine and into the sink, and in this process some of the heat is converted into work.

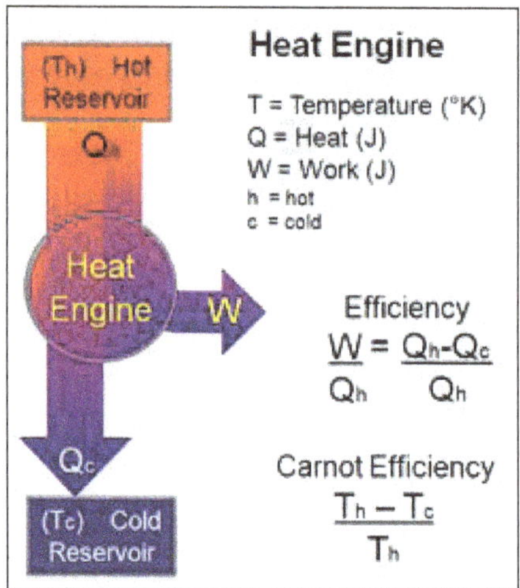

Heat engine theory concerns only the process of converting heat into mechanical energy, not the method of providing the heat, the combustion process. Combustion is a separate conversion process and is subject to its own efficiency losses. In some practical systems such as steam turbines these two processes are physically separate, but in internal combustion engines, which account for the majority of engines, the two processes take place in the same chamber, at the same time.

Efficiency of Heat Engine

The efficiency 'η' of the heat engine is the ratio between its output of work to the heat supply of the heat engine. Let us derive an expression for the efficiency of a heat engine,

$$\eta = W / Q1$$

where, W is the work done by the engine and Q1 is the heat absorbed from the source. After each cycle, the engine returns to its original state so that it does not affect its internal energy:

$$\Delta U = 0$$

$$W = Q1 - Q2$$

Hence the engine efficiency is:

$$\eta = (Q1 - Q2) / Q1$$

$$\eta = 1 - Q2 / Q1$$

So here efficiency will be 100% but in actual, this is not possible because there will be some loss of energy in the system. Hence for every engine, there is a limit for its efficiency. Gasoline and diesel engines, jet engines, and steam turbines that generate electricity are all examples of heat engines.

Example: A Lawn Mower

A lawn mower is rated to have an efficiency of 25% and an average power of 3.00 kW. What are:

- The average work.
- The minimum heat discharge into the air by the lawn mower in one minute of use?

Strategy

From the average power—that is, the rate of work production—we can figure out the work done in the given elapsed time. Then, from the efficiency given, we can figure out the minimum heat discharge $Q_c = Q_h(1-e)$ with $Q_h = Q_c + W$.

Solution:

- The average work delivered by the lawn mower is:

  $$W = P\Delta t$$
  $$= 3.00 \times 10^3 \times 60 \times 1.00\ J$$
  $$= 180\, kJ.$$

- The minimum heat discharged into the air is given by:

  $$Q_c = Q_h(1-e)$$
  $$= (Q_c + W)(1-e)$$

  Which leads to:

  $$Q_c = W(1/e-1)$$
  $$= 180 \times (1/0.25-1)\, kJ = 540\ kJ.$$

Significance

As the efficiency rises, the minimum heat discharged falls. This helps our environment and atmosphere by not having as much waste heat expelled.

Carnot Cycle

Carnot cycle is an ideal cyclical sequence of changes of pressures and temperatures of a fluid, such as a gas used in an engine, conceived early in the 19th century by the French engineer Sadi Carnot. It is used as a standard of performance of all heat engines operating between a high and a low temperature.

The Carnot cycle consists of the following four processes:

- A reversible isothermal gas expansion process. In this process, the ideal gas in the system absorbs q_{in} amount heat from a heat source at a high temperature T_h, expands and does work on surroundings.
- A reversible adiabatic gas expansion process. In this process, the system is thermally insulated. The gas continues to expand and do work on surroundings, which causes the system to cool to a lower temperature, T_l.
- A reversible isothermal gas compression process. In this process, surroundings do work to the gas at T_l, and causes a loss of heat, q_{out}.

- A reversible adiabatic gas compression process. In this process, the system is thermally insulated. Surroundings continue to do work to the gas, which causes the temperature to rise back to T_h.

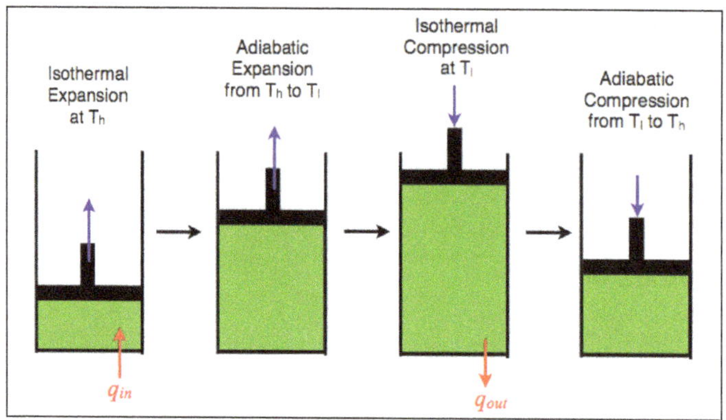

An ideal gas-piston model of the Carnot cycle.

P-V Diagram

The P-V diagram of the Carnot cycle is shown in figure. In isothermal processes I and III, $\Delta U = 0$ because $\Delta T = 0$. In adiabatic processes II and IV, $q = 0$. Work, heat, ΔU, and ΔH of each process in the Carnot cycle are summarized in table.

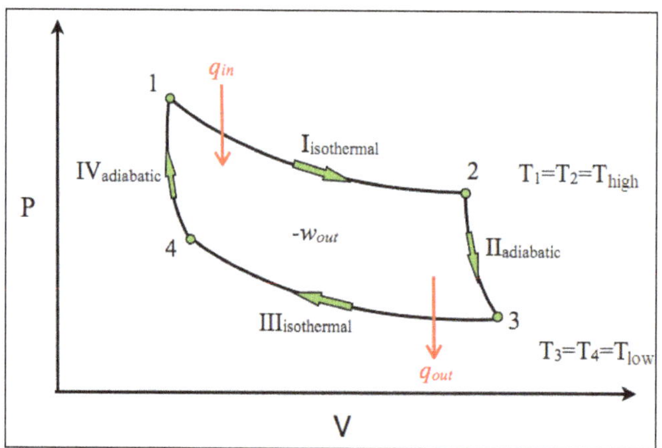

A P-V diagram of the Carnot Cycle.

Table: Work, heat, ΔU, and ΔH in the P-V diagram of the Carnot Cycle.

Process	w	q	ΔU	ΔH
I	$-nRT_{high} \ln\left(\frac{V_2}{V_1}\right)$	$nRT_{high} \ln\left(\frac{V_2}{V_1}\right)$	0	0
II	$n\bar{C}_v\left(T_{low} - T_{high}\right)$	0	$n\bar{C}_v\left(T_{low} - T_{high}\right)$	$n\bar{C}_p\left(T_{low} - T_{high}\right)$
III	$-nRT_{low} \ln\left(\frac{V_4}{V_3}\right)$	$nRT_{low} \ln\left(\frac{V_4}{V_3}\right)$	0	0

IV	$n\bar{C}_v\left(T_{high}-T_{low}\right)$	0	$n\bar{C}_v\left(T_{hight}-T_{low}\right)$	$n\bar{C}_p\left(T_{high}-T_{low}\right)$
Full Cycle	$-nRT_{high}\ \text{In}\left(\frac{V_2}{V_1}\right)$ $-nRT_{low}\ \text{In}\left(\frac{V_4}{V_3}\right)$	$nRT_{high}\ \text{In}\left(\frac{V_2}{V_1}\right)$ $+nRT_{low}\ \text{In}\left(\frac{V_4}{V_3}\right)$	0	0

T-S Diagram

The T-S diagram of the Carnot cycle is shown in figure. In isothermal processes I and III, ΔT = 0. In adiabatic processes II and IV, ΔS = 0 because dq = 0. ΔT and ΔS of each process in the Carnot cycle are shown in table.

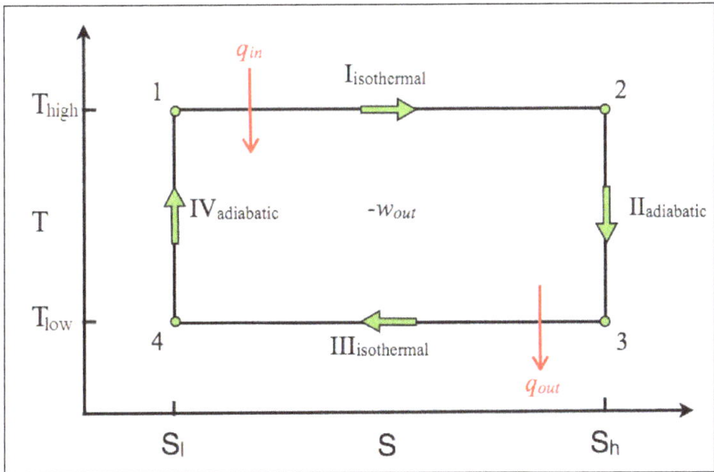

A T-S diagram of the Carnot Cycle.

Table: Work, heat, and ΔU in the T-S diagram of the Carnot Cycle.

Process	ΔT	ΔS
I	0	$-nR\ \text{In}\left(\frac{V_2}{V_1}\right)$
II	$T_{low}-T_{high}$	0
III	0	$-nR\ \text{In}\left(\frac{V_4}{V_3}\right)$
IV	$T_{high}-T_{low}$	0
Full Cycle	0	0

Efficiency

The Carnot cycle is the most efficient engine possible based on the assumption of the absence of incidental wasteful processes such as friction, and the assumption of no conduction of heat between different parts of the engine at different temperatures. The efficiency of the carnot engine is defined as the ratio of the energy output to the energy input.

$$\text{efficiency} = \frac{\text{net work done by heat engine}}{\text{heat absorbed by heat engine}} = \frac{-w_{sys}}{q_{high}}$$

$$= \frac{nRT_{high}\,\text{In}\left(\frac{V_2}{V_1}\right) + nRT_{low}\,\text{In}\left(\frac{V_4}{V_3}\right)}{nRT_{high}\,\text{In}\left(\frac{V_2}{V_1}\right)}$$

Since processes II (2-3) and IV (4-1) are adiabatic,

$$\left(\frac{T_2}{T_3}\right)^{Cv/R} = \frac{V_3}{V_2}$$

and

$$\left(\frac{T_1}{T_4}\right)^{Cv/R} = \frac{V_4}{V_1}$$

And since $T_1 = T_2$ and $T_3 = T_4$,

$$\frac{V_3}{V_4} = \frac{V_2}{V_1}$$

Therefore,

$$\text{efficiency} = \frac{nRT_{high}\,\text{In}\left(\frac{V_2}{V_1}\right) - nRT_{low}\,\text{In}\left(\frac{V_2}{V_1}\right)}{nRT_{high}\,\text{In}\left(\frac{V_2}{V_1}\right)}$$

$$\text{efficiency} = \frac{T_{high} - T_{low}}{T_{high}}$$

Carnot Theorem

Carnot's theorem also known as Carnot's rule was developed by Nicolas Léonard Sadi Carnot in the year 1824, with the principle that there are limits on maximum efficiency for any given heat engine. It depends mainly on hot and cold reservoir temperatures.

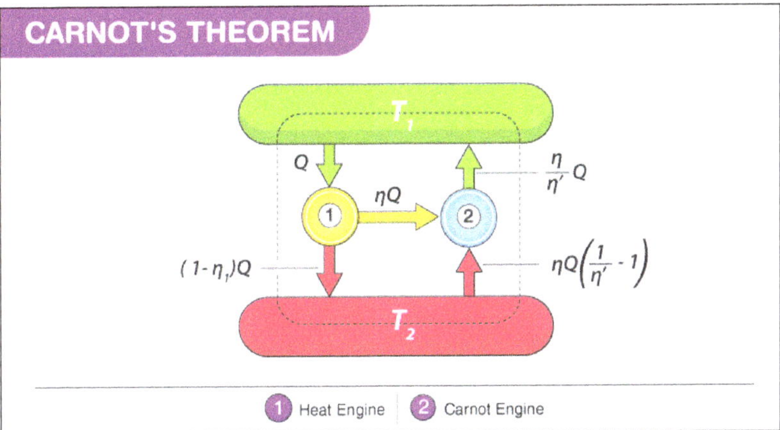

Carnot's theorem states that:

- Heat engines that are working between two heat reservoirs are less efficient than the Carnot heat engine that are operating between same reservoirs.
- Irrespective of the operation details, every Carnot engine is efficient between two heat reservoirs.
- Maximum efficiency is given as:

$$\eta_{\max} = \eta_{Carnot} = 1 - \frac{T_c}{T_H}$$

where,

- TC: absolute temperature of cold reservoir.
- TH: absolute temperature of hot reservoir.
- η: ratio of work done by the engine to heat drawn out of the hot reservoir.

Processes involved in thermodynamics can be carried out in the following two ways:

- Reversible Engine:

  The efficiency of all reversible engines remains same that work between two same heat reservoirs.

$$\Delta S = \int_a^b \frac{dQ_{rev}}{T}$$

where,

ΔS: change in entropy

T: temperature

$\int_a^b$: to show that this is a path function

- Irreversible engine:

  There is no irreversible engine that is more efficient than Carnot engine working between two same reservoirs.

Example of an irreversible engine are:

- Plastic deformation.
- Friction.
- Spontaneous chemical reaction.

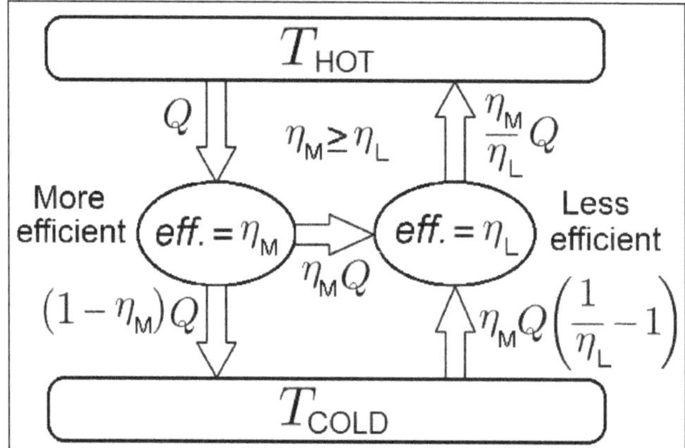

An impossible situation: A heat engine cannot drive a less efficient (reversible) heat engine without violating the second law of thermodynamics.

The proof of the Carnot theorem is a proof by contradiction, or reductio ad absurdum, as illustrated by the figure showing two heat engines operating between two reservoirs of different temperature. The heat engine with more efficiency (η_M) is driving a heat engine with less efficiency (η_L), causing the latter to act as a heat pump. This pair of engines receives no outside energy, and operates solely on the energy released when heat is transferred from the hot and into the cold reservoir. However, if $\eta_M > \eta_L$, then the net heat flow would be backwards, i.e., into the hot reservoir:

$$Q_{\text{hot}}^{\text{out}} = Q < \frac{\eta_M}{\eta_L} Q = Q_{\text{hot}}^{\text{in}}.$$

It is generally agreed that this is impossible because it violates the second law of thermodynamics.

We begin by verifying the values of work and heat flow depicted in the figure. First, we must point out an important caveat: the engine with less efficiency (η_L) is being driven as a heat pump, and therefore must be a *reversible* engine. If the less efficient engine (η_L) is not reversible, then the device could be built, but the expressions for work and heat flow shown in the figure would not be valid.

By restricting our discussion to cases where engine (η_L) has less efficiency than engine (η_M), we are able to simplify notation by adopting the convention that all symbols, Q and W represent non-negative quantities (since the direction of energy flow never changes sign in all cases where

$\eta_L \le \eta_M$). Conservation of energy demands that for each engine, the energy which enters, E_{in}, must equal the energy which exits, E_{out}:

$$E_{\text{in}}^{M} = Q = (1-\eta_M)Q + \eta_M Q = E_{\text{out}}^{M}$$

$$E_{\text{in}}^{L} = \eta_M Q + \eta_M Q\left(\frac{1}{\eta_L} - 1\right) = \frac{\eta_M}{\eta_L} Q = E_{\text{out}}^{L}$$

The figure is also consistent with the definition of efficiency as $\eta = W / Q_h$ for both engines:

$$\eta_M = \frac{W_M}{Q_h^M} = \frac{\eta_M Q}{Q} = \eta_M$$

$$\eta_L = \frac{W_L}{Q_h^L} = \frac{\eta_M Q}{\frac{\eta_M}{\eta_L} Q} = \eta_L$$

It may seem odd that a hypothetical heat pump with low efficiency is being used to violate the second law of thermodynamics, but the figure of merit for refrigerator units is not efficiency, W / Q_h, but the coefficient of performance (COP), which is Q_c / W. A reversible heat engine with low thermodynamic efficiency, W / Q_h delivers more heat to the hot reservoir for a given amount of work when it is being driven as a heat pump.

Having established that the heat flow values shown in the figure are correct, Carnot's theorem may be proven for irreversible and the reversible heat engines.

Reversible Engines

To see that every reversible engine operating between reservoirs T_1 and T_2 must have the same efficiency, assume that two reversible heat engines have different values of η, and let the more efficient engine (M) drive the less efficient engine (L) as a heat pump. As the figure shows, this will cause heat to flow from the cold to the hot reservoir without any external work or energy, which violates the second law of thermodynamics. Therefore both (reversible) heat engines have the same efficiency, and we conclude that:

> All reversible engines that operate between the same two heat reservoirs have the same efficiency.

This is an important result because it helps establish the Clausius theorem, which implies that the change in entropy is unique for all reversible processes,

$$\Delta S = \int_a^b \frac{dQ_{\text{rev}}}{T}$$

Over all paths (from a to b in V-T space). If this integral were not path independent, then entropy, S, would lose its status as a state variable.

Irreversible Engines

If one of the engines is irreversible, it must be the (M) engine, placed so that it reverse drives the less efficient but reversible (L) engine. But if this irreversible engine is more efficient than the reversible engine, (i.e., if $\eta_M > \eta_L$), then the second law of thermodynamics is violated. And, since the Carnot cycle represents a reversible engine, we have the first part of Carnot's theorem:

> No irreversible engine is more efficient than the Carnot engine operating between the same two reservoirs.

Applications of Carnot's Theorem

- Carnot's theorem finds application in engines that convert thermal energy to work.
- Refrigeration: Method of removal of heat from the at low temperature and dissipating it to a higher temperature. This is a reversible process.

Carnot Engine

Carnot engine is a theoretical thermodynamic cycle proposed by Nicolas Léonard Sadi Carnot in 1824. Carnot states that a hot body is required that generates heat and a cold body to which the caloric is conveyed, which produces a mechanical work in the process. It also states that said work is free of the material that is used to create heat and the construction and design material of the machine.

Modern Diagram

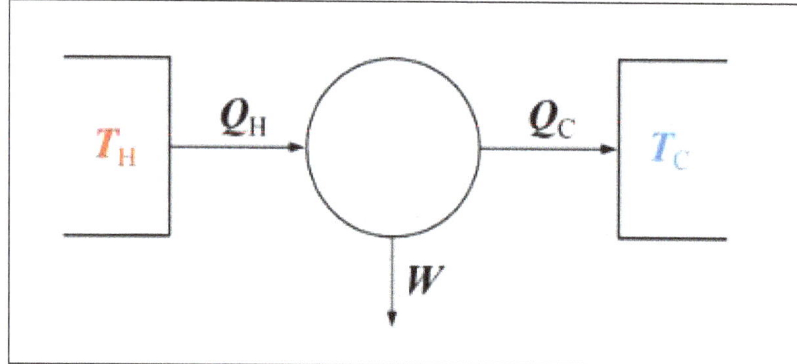

The figure displays a block diagram of a general heat engine, like the Carnot engine. In the diagram, the "working body", a word presented by Clausius in 1850, can be any vapor or fluid body through which heat "Q" can be transmitted to yield work. Carnot had proposed that the fluid body could be any material capable of expansion, such as vapor of alcohol, the vapor of mercury, the vapor of water, a permanent air or gas etc. While, in these initial years, engines came in a number of patterns, usually Q_H was delivered by a boiler, wherein water was boiled over a heater; Q_C was usually delivered by a stream of cold flowing water in the form of a condenser situated on a separate part of the engine. The yield work, W, denotes the movement of the piston as it is used to rotate a crank-arm, which in turn was normally used to power a pulley so as to lift water out of submerged salt mines. Carnot states work as "weight lifted through a height".

Carnot Engine Principles

Carnot principles are just for cyclical devices such as heat engines, which state that:

- The effectiveness of an irreversible heat engine is always less than the efficiency of a reversible one functioning between the similar two reservoirs.
- The effectiveness of all reversible heat engines working between the similar two reservoirs is equal.

To escalate the thermal efficiency of a gas power turbine, it is essential to increase the temperature of the combustion room. such as, turbines blades cannot hold out the high temperature gas and which will eventually lead to early fatigue.

Carnot's Theorem

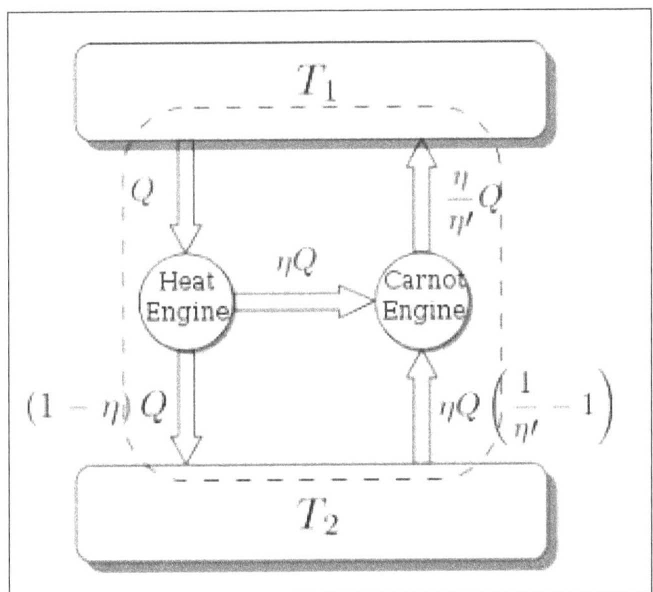

This theorem denes that no engine functioning between two known temperatures can be more effective than a reversible engine functioning between the similar two temperatures and that all the reversible engines functioning between the same two temperatures have the similar efficiency, whatever the working material might be. As per the Carnot theorem, the reversible engine will always have greater productivity than the irreversible one. The reversible heat engine engine works on a reverse cycle and behaves as a heat pump.

The Efficiency of Carnot's Cycle

The Carnot cycle is reversible signifying the upper limit on the efficiency of an engine cycle. Practical engine cycles are irreversible and therefore have inherently much lower efficiency than the Carnot efficiency when working at similar temperatures. One of the factors determining efficiency is the addition of the working fluid in the cycle and its removal. The Carnot cycle reaches maximum efficiency because all the heat is pushed to the working fluid at the maximum temperature.

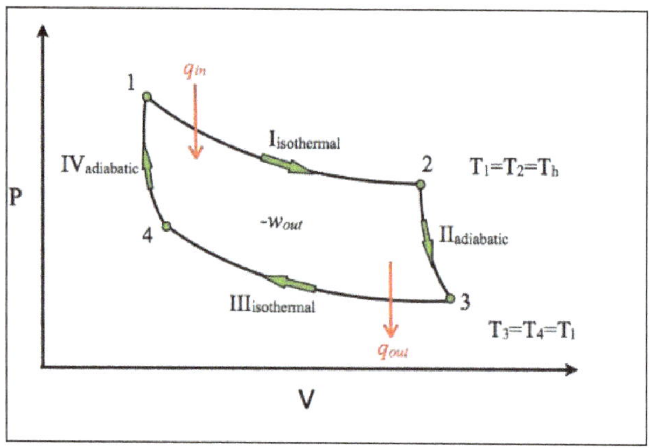

Steps in Cycle

Step 1: Isothermal Expansion

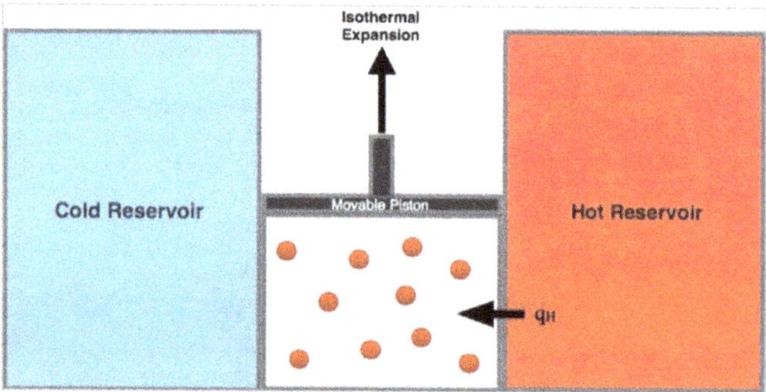

Heat is passed reversibly from the high temperature pool at fixed temperature T_H (isothermal heat absorption). In this step the gas is allowed to expand, doing work on the surroundings by pushing up the piston. Even though the pressure drops from points 1 to 2 the temperature of the gas does not alter in the process because it is in thermal contact with the hot pool at T_h, and therefore the expansion is isothermal. Heat energy Q_1 is absorbed from the high temperature pool resulting in an increase in the entropy of the gas by the amount.

$$\Delta S_1 = Q_1 / T_h$$

Step 2: Isentropic (Reversible Adiabatic)

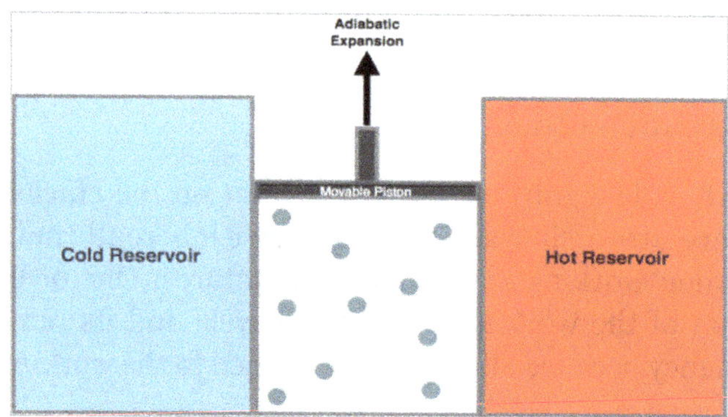

Expansion of the gas. In this step the gas in the engine is thermally shielded from both the hot and cold pools. Thus they cannot gain nor lose heat, so it called an 'adiabatic' process. The gas rises to expand by the drop in pressure, doing work on the surroundings (raising the piston), and losing a volume of internal energy similar to the work done. The gas begins to expansion without heat input causes it to cool to the "cold" temperature, Tc. The entropy remains the same.

Step 3: Isothermal Compression

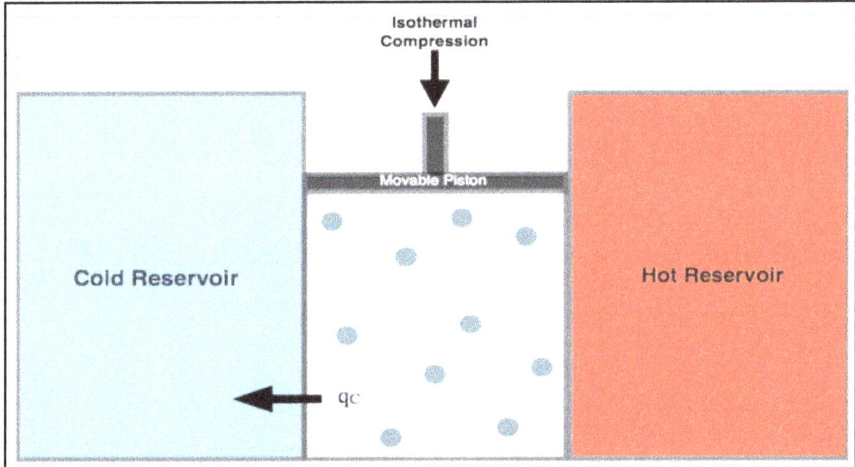

Heat shifted reversibly to a low-temperature pool at constant temperature TC. (isothermal heat elimination) Now the gas in the engine is in thermal contact with the cold pool at temperature Tc. The surroundings do work on the gas, by pushing the piston down, causing a volume of heat energy Q2 to leave the system to a low-temperature pool and the entropy of the system to decline by the amount. (This is the equal amount of entropy absorbed in step 1, as can be observed from the Clausius inequality.)

Step 4: Adiabatic Reversible Compression

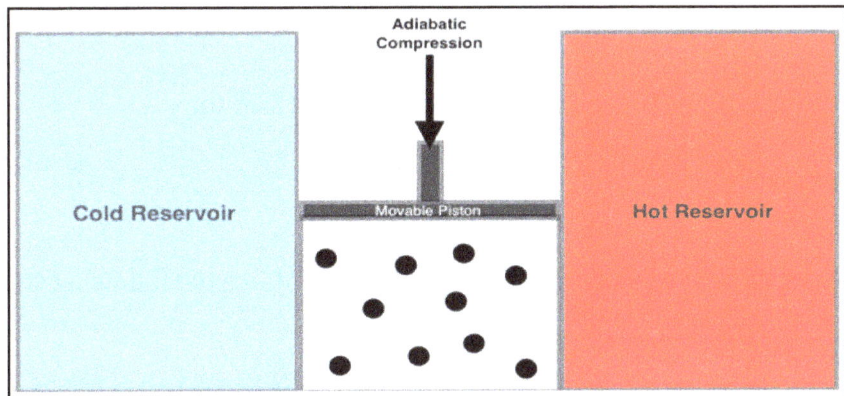

Once again the gas in the engine is thermally shielded from the hot and cold pools, and the engine is expected to be frictionless, therefore it is reversible. In this step, the surroundings do work on the gas, pushing the piston down more, rising it's internal energy, compressing it, and producing its temperature to rise back to Th due only to the work added to the system, but the entropy remains the same. At this stage, the gas is in the same state as at the beginning of the step.

Therefore, the total work done by the gas on the environment in one complete cycle is shown by:

$$W = W_{1\to2} + W_{2\to3} + W_{3\to4} + W_{4\to1}$$

$$W = \mu\ RT_1\, In\frac{v_2}{v_1} - \mu RT_2\, In\frac{v_3}{v_4}$$

$$\text{Net efficiency} = \frac{\text{Net workdone by the gas}}{\text{Heat absorbed by the gas}}$$

$$\text{Net efficiency} = \frac{W}{Q_1} = \frac{Q_1 - Q_2}{Q_1} = 1 - \frac{Q_2}{Q_1} = 1 - \frac{T_2}{T_1}\frac{In\frac{v_3}{v_4}}{In\frac{v_2}{v_1}}$$

As step 2–>3 is an adiabatic process, we can write $T_1 V_2^{Y-1} = T_2 V_3^{Y-1}$

or

$$\frac{v_2}{v_3} = \left(\frac{T_2}{T_1}\right)^{\frac{1}{\gamma-1}}$$

Equally, for process 4–>1, we can write:

$$\frac{v_1}{v_2} = \left(\frac{T_2}{T_1}\right)^{\frac{1}{\gamma-1}}$$

This indicates,

$$\frac{v_2}{v_3} = \frac{v_1}{v_2}$$

So, the expression for a net efficiency of Carnot engine decreases to:

$$\text{Net efficiency} = 1 - \frac{T_2}{T_1}$$

The Carnot engine cycle when behaving as a heat engine contains the following steps:

- Isentropic compression of the gas.
- Reversible or changeable adiabatic (Isentropic) expansion of the gas.
- Reversible or changeable isothermal expansion of the gas at the "hot" temperature.

Applications of the Carnot Cycle

Thermal machines or thermal devices are one of the applications of the Carnot cycle. The heat pumps to generate heating, the refrigerators to yield cooling, the steam turbines used in the ships, the combustion engines of the combustion vehicles and the reaction turbines of an airplane are some of the examples that we can give.

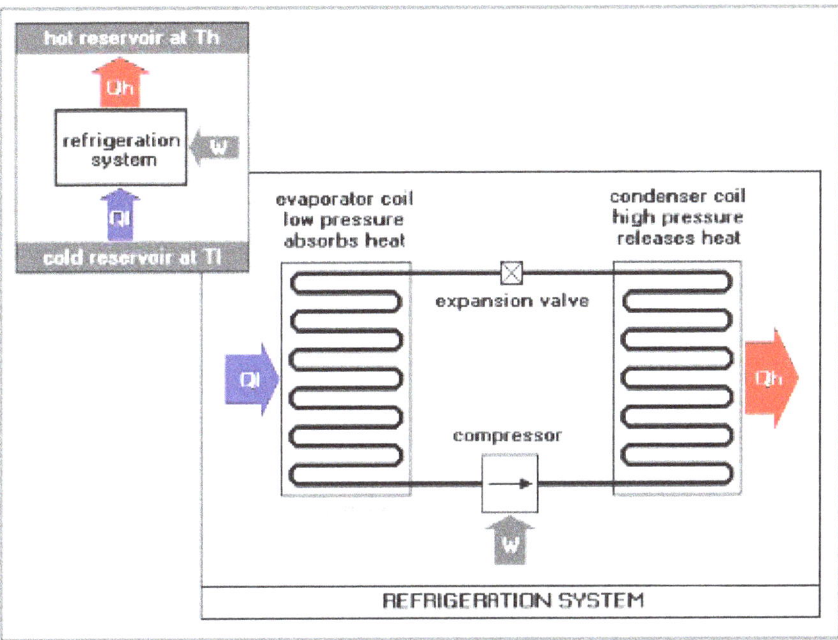

Limitations

This equation shows that the bigger the temperature range, the more effective is the cycle is.

- T3: In practice, T3 cannot be decreased below about 300 K (27 °C), equivalent to a condenser pressure of 0.035 bar. This is because of two tractors:
    - Condensation of steam needs a bulk supply of cooling water and such a continuous natural supply below the atmospheric temperature of around 15 °C is unavailable.
    - If the condenser is to be of a rational size and cost, the temperature difference between the condensing steam and the cooling water must be minimum 10 °C.
- TI: The extreme cycle temperature T1 is also limited to near 900 K (627 °C) by the strength of the substance available for the extremely stressed parts of the plant, like boiler tubes and turbine blades. This upper limit is known as the metallurgical limit.
- Critical Point: The steam of the Carnot cycle has a tremendous cycle temperature of well beneath this metallurgical limit due to the properties of steam; it is limited to the critical-point temperature of 374°C (647 K). Therefore modern substance cannot be used to their best advantage with this cycle when steam is the functioning fluid. Also, because the saturated steam and water curves converge to the critical point, a plant working on the Carnot cycle with its extreme temperature near the critical-point temperature would have a very large s.s.c., i.e. it would be very big in size and very costly.

- Compression Process: Compressing a very wet steam blend would need a compressor of cost and size of comparable with the turbine. It would absorb work comparable with the advanced by the turbine. It would have a short life because of blade corrosion and cavitation problem. These reasons the Carnot cycle is not useful.

Refrigerator

A refrigerator is a device which is designed to remove heat from a space that is at lower temperature than its surroundings. The same device can be used to heat a volume that is at higher temperature than the surroundings. In this case the device is called a Heat Pump. The distinction between a refrigerator and a heat pump is one of purpose rather than principle.

The Clausius statement of the Second Law of Thermodynamics asserts that it is impossible to construct a device that, operating in a cycle, has no effect other than the transfer of heat from a cooler to a hotter body. This means that energy will not flow from cold to hot regions without outside assistance. The refrigerator and heat pump both satisfy the Clausius requirement of external action through the application of mechanical power or equivalent natural transfers of heat.

Continuous refrigeration can be achieved by several processes. Effectively any heat engine cycle, when reversed, becomes a refrigeration cycle. The vapor compression cycle is the most commonly used in refrigeration and air condition applications. The vapor absorption cycle provides an alternative system, particularly in applications where heat is economically available. Steam-jet systems are also being successfully used in many cooling applications while air-cycle refrigeration is often used for aircraft cooling.

Reversed Heat Engine Cycles

Mechanical refrigeration processes, of which the vapor compression cycle is an example, belong to the general class of reversed heat engine cycles. The figure represents, schematically, the extraction of heat at rate $\dot{Q}$ from a cold body at temperature T_C. The process requires the expenditure of work W and the sum $\left(\dot{Q}+\dot{W}\right)$ is discharged at a higher temperature T_H.

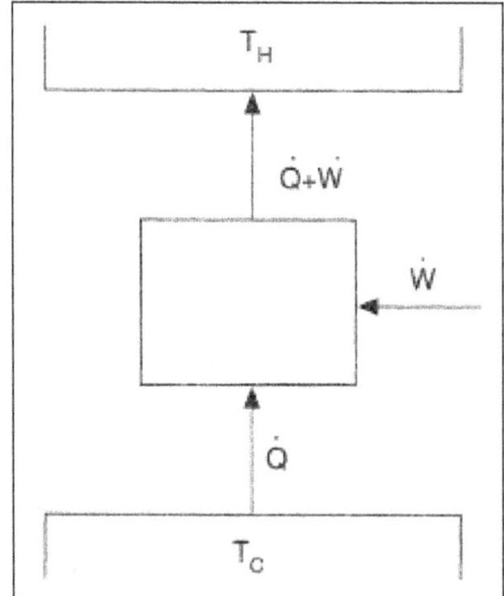

Reversed heat engine cycle.

The ideal cycle against which any practical reversed heat engine may be compared is the reversible or Carnot Cycle for which, in accordance with the Second Law of Thermodynamics, the following relationship applies:

$$\dot{W} = \frac{(T_H - T_C)}{T_C}\dot{Q}$$

One measure of the efficiency of such a process is given by:

$$\frac{\dot{W}}{\dot{Q}} = \frac{T_H - T_C}{T_C}$$

Clearly the smaller the value of the ratio $\dot{Q} / \dot{W}o$ the more efficient is the process.

It is more usual to describe the efficiency of a reversed heat engine by the inverse of this ratio, known as the Coefficient Of Performance (COP):

$$COP = \frac{\dot{Q}}{\dot{W}} = \frac{T_C}{T_H - T_C}$$

It will be observed that the COP may be greater than unity and that it becomes greater as the temperature difference decreases. A real refrigerator or reversed heat engine will have a COP less than that of the ideal Carnot Cycle engine as given by the above equation.

The reversed Carnot Cycle is represented on the Temperature-Entropy (T-S) diagram by a rectangle, and is composed of four reversible processes;

- 4-1 isothermal expansion, during which heat (the refrigeration load) flows from the cold space to the working fluid.
- 1-2 adiabatic compression.
- 2-3 isothermal compression in which heat flows from refrigerant to the hot space.
- 3-4 adiabatic expansion.

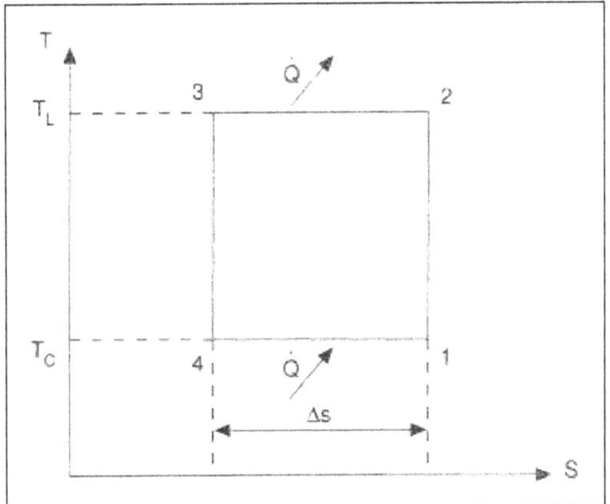

Temperature-entropy diagram for ideal reversed Carnot Cycle.

Basic Vapor Compression Cycle and Components

Vapor compression refrigeration, as the name suggests, employs a compression process to raise the pressure of a working fluid vapor (refrigerant) flowing from an evaporator at low pressure P_L to a high pressure P_H as shown in figure. The refrigerant then flows through a condenser at the higher pressure P_H, through a throttling device, and back to the low pressure, P_L in the evaporator. The pressures P_L and P_H correspond to the refrigerant saturation temperatures, T_1 and T_5 respectively.

The T-S diagram for this real cycle, is somewhat different from the rectangular shape of the Carnot Cycle.

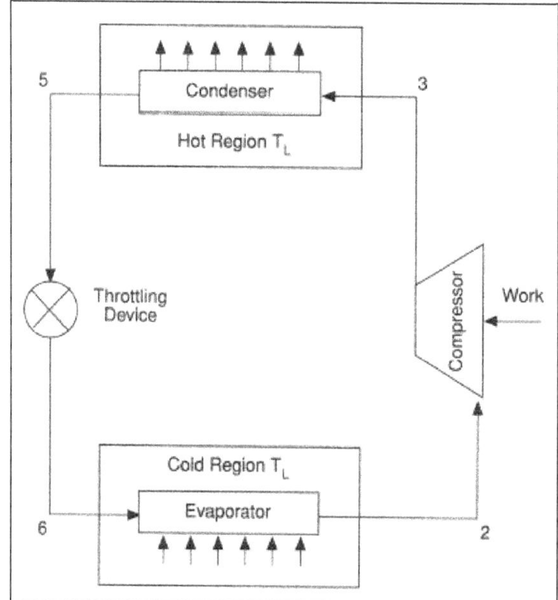

Basic vapor compression refrigerator.

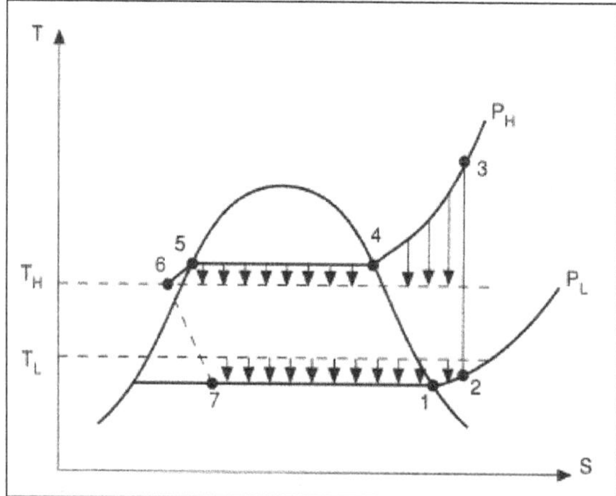

T-S diagram for basic vapor compression cycle.

The cycle processes can be described as follows:

- 7-1 Evaporation of the liquefied refrigerant at constant temperature $T_1 = T_7$.
- 1-2 Superheating of the vapor from temperature T_1 to T_2 at constant pressure P_L.

- 2-3 Compression (not necessarily adiabatic) from temperature T_2 and pressure P_L to temperature T_3 and pressure P_H.
- 3-4 Cooling of the super-heated vapor to the saturation temperature T_4.
- 4-5 Condensation of the vapor at temperature $T_4 = T_5$ and pressure P_H.
- 5-6 Subcooling of the liquid from T_5 to T_6 at pressure P_H.
- 6-7 Expansion from pressure P_H to pressure P_L at constant enthalpy.

A further difference between the real cycle and the ideal is that temperature T_1 at which evaporation takes place is lower than the temperature T_L of the cold region so heat transfer can take place. Similarly the temperature T_4 of the heat rejection must be higher than the hot region temperature T_H to bring about heat transfer in the condenser.

It is usual for the vapor-compression cycle to be plotted on a pressure-enthalpy (p-h) diagram as shown in figure.

Refrigerants

Refrigerants are the working fluids in refrigeration systems. They must have certain characteristics which include good refrigeration performance, low flammability and toxicity, compatibility with compressor lubricating oils and metals, and good heat transfer properties. They are usually identified by a number that relates to their molecular composition.

In recent years, environmental concerns over the use of chlorofluorocarbons (CFCs) as the working fluids in refrigeration and air-conditioning plants have led to the development of alternative fluids. The majority of these fall into two categories, hydrofluorocarbons (HDCs) which contain no chlorine and have zero ozone depletion potential and hydrochlorofluorocarbons (HCFCs), which do contain chlorine, but the addition of hydrogen to the CFC structure allows virtually all the chlorine to be dispersed in the lower atmosphere before it can reach the ozone layer. HCFCs therefore have much lower ozone depletion potentials, ranging from 2 to 10% that of CFCs.

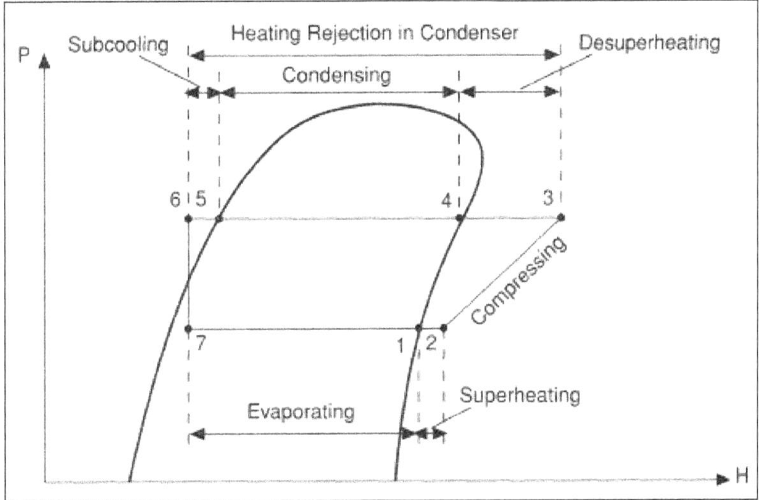

P-H representation of vapor compression cycles.

Vapor-absorption Cycles

Recently interest has been increasing in these cycles because of their potential use as part of energy-saving plants and also because they use more environmentally friendly refrigerants than vapor-compression cycles. A basic vapor-absorption system is shown schematically in figure. The condenser, throttling valve and evaporator are essentially the same as in the vapor compression system. The major difference is the replacement of the compressor with an absorber, a generator, and a solution pump. A second throttling valve is also used to maintain the pressure difference between the absorber (at the evaporator pressure) and the generator (at the condenser pressure).

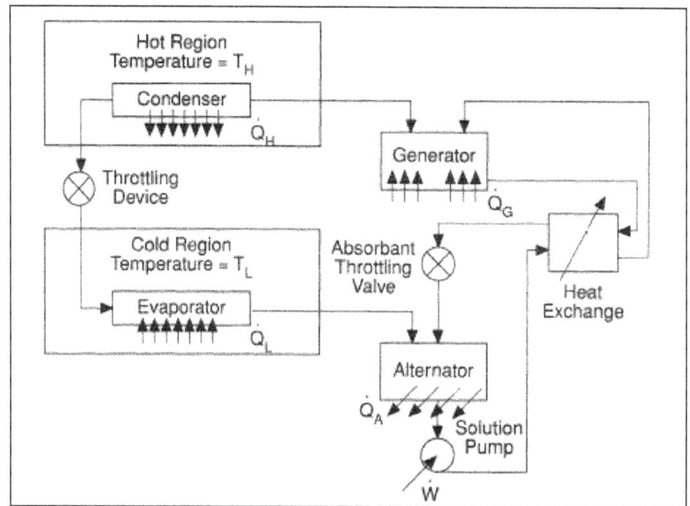

Basic absorption refrigerant system.

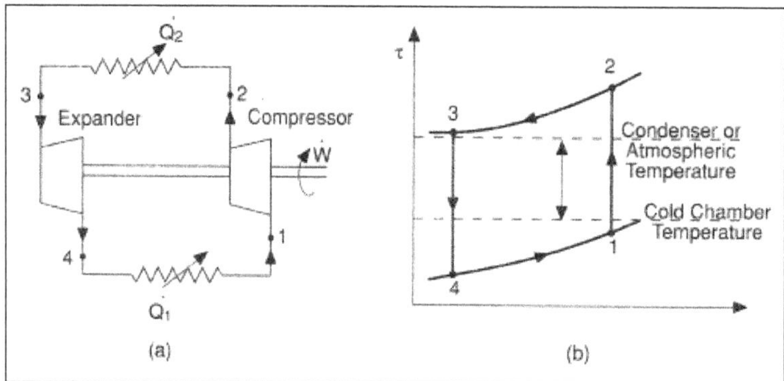

Simple gas-cycle refrigerant system.

The refrigerant on leaving the evaporator is absorbed in a low-temperature absorbing medium, some heat, Q_A, being rejected in the process. The refrigerant-absorbent solution is then pumped to the higher pressure and is heated in the generator, Q_G. Refrigerants vapor then separates from the solution due to the high pressure and temperature in the generator. The vapor passes to the condenser and the weak solution is throttled back to the absorber. A heat exchanger may be placed between the absorber and the generator to increase the energy efficiency of the system. The work done in pumping the liquid solution is much less than that required by the compressor in the equivalent vapor-compression cycle. The main energy input to the system, Q_G, may be supplied in any convenient form such as a fuel burning device, electrical heating, steam, solar energy or waste heat. Appropriate refrigerant/absorbent combinations must be selected. One common combination uses ammonia as refrigerant and water as absorbent. An alternative combination is water as refrigerant and lithium bromide as absorbent.

Gas-cycle Refrigeration

Gas-cycle refrigeration, is essentially, a reversed Joule cycle (gas turbine cycle). As the name indicates, the refrigerant in these systems is a gas. The system, as shown in figure, is basically the same as that of the vapor-compression cycle. The main difference is the replacement of the throttling valve by an expander.

The cycle can be described as follows:

- 1-2 Adiabatic compression.
- 2-3 Constant pressure cooling.
- 3-4 Adiabatic expansion.
- 4-1 Constant pressure heating (cooling effect).

As can be seen from figure above , the gas does not receive and reject heat at constant temperature, and, therefore, the gas cycle is less efficient than the vapor cycle for given evaporator and condenser temperatures. Gas-cycle systems are mostly used in air conditioning applications where the working fluid-air can be ejected at T_4. A common application is in the air conditioning of aircraft. Air, held from the engine compressor, is cooled in a heat exchanger and then expanded through a turbine. The power from the turbine is used to drive a fan which provides the cooling air for the heat exchanger. Air at T_4 is ejected into the cabin to provide the required cooling.

Third Law of Thermodynamics

The third law of thermodynamics states that the entropy of a perfect crystal at a temperature of zero Kelvin (absolute zero) is equal to zero.

Entropy, denoted by 'S', is a measure of the disorder/randomness in a closed system. It is directly related to the number of microstates (a fixed microscopic state that can be occupied by a system) accessible by the system, i.e. the greater the number of microstates the closed system can occupy, the greater its entropy. The microstate in which the energy of the system is at its minimum is called the ground state of the system.

At a temperature of zero Kelvin, the following phenomena can be observed in a closed system:

- The system does not contain any heat.
- All the atoms and molecules in the system are at their lowest energy points.

Therefore, a system at absolute zero has only one accessible microstate – it's ground state. As per the third law of thermodynamics, the entropy of such a system is exactly zero.

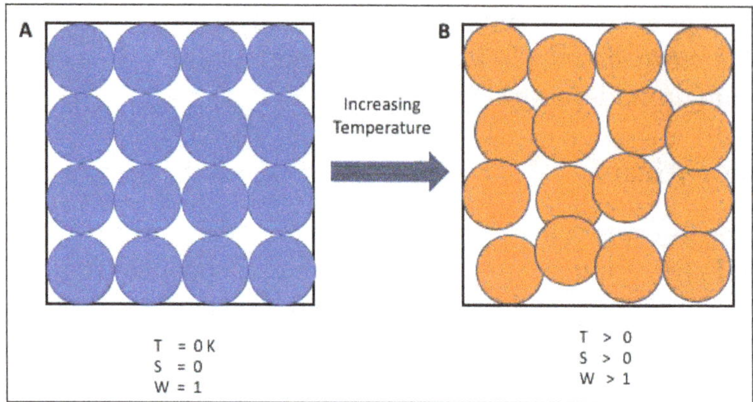

This law was developed by the German chemist Walther Nernst between the years 1906 and 1912.

Alternate Statements of the 3rd Law of Thermodynamics

The Nernst statement of the third law of thermodynamics implies that it is not possible for a process to bring the entropy of a given system to zero in a finite number of operations.

The American physical chemists Merle Randall and Gilbert Lewis stated this law in a different manner: when the entropy of each and every element (in their perfectly crystalline states) is taken as 0 at absolute zero temperature, the entropy of every substance must have a positive, finite value. However, the entropy at absolute zero can be equal to zero, as is the case when a perfect crystal is considered.

The Nernst-Simon statement of the 3rd law of thermodynamics can be written as: for a condensed system undergoing an isothermal process that is reversible in nature, the associated entropy change approaches zero as the associated temperature approaches zero.

Another implication of the third law of thermodynamics is: the exchange of energy between two thermodynamic systems (whose composite constitutes an isolated system) is bounded.

Why is it Impossible to Achieve a Temperature of Zero Kelvin?

For an isentropic process that reduces the temperature of some substance by modifying some parameter X to bring about a change from 'X_2' to 'X_1', an infinite number of steps must be performed in order to cool the substance to zero Kelvin.

This is because the third law of thermodynamics states that the entropy change at absolute zero temperatures is zero. The entropy v/s temperature graph for any isentropic process attempting to cool a substance to absolute zero is illustrated below.

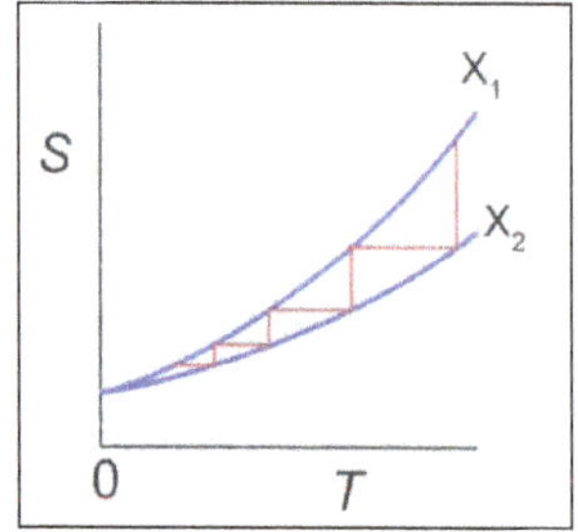

From the graph, it can be observed that – the lower the temperature associated with the substance, the greater the number of steps required to cool the substance further. As the temperature approaches zero kelvin, the number of steps required to cool the substance further approaches infinity.

Mathematical Explanation of the Third Law

As per statistical mechanics, the entropy of a system can be expressed via the following equation:

$$S - S_0 = k_B \ln \Omega$$

where,

- S is the entropy of the system.
- S_0 is the initial entropy.
- k_B denotes the Boltzmann constant.
- Ω refers to the total number of microstates that are consistent with the system's macroscopic configuration.

Now, for a perfect crystal that has exactly one unique ground state, $\Omega = 1$. Therefore, the equation can be rewritten as follows:

$$S - S_0 = k_B \ln(1) = 0 \text{ [because } \ln(1) = 0\text{]}$$

When the initial entropy of the system is selected as zero, the following value of 'S' can be obtained:

$$S - 0 = 0 \Rightarrow S = 0$$

Thus, the entropy of a perfect crystal at absolute zero is zero.

Applications of the Third Law of Thermodynamics

An important application of the third law of thermodynamics is that it helps in the calculation of the absolute entropy of a substance at any temperature 'T'. These determinations are based on the heat capacity measurements of the substance. For any solid, let S_0 be the entropy at 0 K and S be the entropy at T K, then:

$$\Delta S = S - S_0 = \int_0^T \frac{C_p dT}{T}$$

According to the third law of thermodynamics, $S_0 = 0$ at 0 K,

$$S = \int_0^T \frac{C_p}{T} dT$$

The value of this integral can be obtained by plotting the graph of C_p / T versus T and then finding the area of this curve from 0 to T. The simplified expression for the absolute entropy of a solid at temperature T is as follows:

$$S = \int_0^T \frac{C_p}{T} dT = \int_0^T C_p \, d \text{ In T}$$

$$= C_p \text{ In } T = 2.303 \, C_p \log \text{T}$$

Here C_p is the heat capacity of the substance at constant pressure and this value is assumed to be constant in the range of 0 to T K.

References

- The-zeroth-law-of-thermodynamics, boundless-physics: courses.lumenlearning.com, Retrieved 21 June, 2019
- Thermal-equilibrium-and-zeroth-law-of-thermodynamics: byjus.com, Retrieved 25 July, 2019
- Zeroth-law-of-thermodynamics-4177952: thoughtco.com, Retrieved 13 August, 2019
- First-law-of-thermodynamics, thermodynamics: toppr.com, Retrieved 15 January, 2019
- Reversible-and-irreversible-process, thermodynamics: toppr.com, Retrieved 17 May, 2019
- Carnot-cycle: britannica.com, Retrieved 17 April, 2019

3

Thermodynamic Processes

There are different types of thermodynamic processes that affect pressure, volume, temperature and heat transfer. These processes are isothermal process, adiabatic process, isochoric process, isobaric process and reversible process. This chapter closely examines these various thermodynamic processes to provide an extensive understanding of the subject.

Thermodynamic state of a system is characterized by its parameters P, V and T. A change in one or more parameters results in a change in the state of the system.

A process by which one or more parameters of thermodynamic system undergo a change is called a thermodynamic process or a thermodynamic change.

The initial and final states are the defining elements of the process. During such a process, a system starts from an initial state i, described by a pressure p_i, volume V_i and a temperature T_i, passes through various quasistatic states to a final state f, described by a pressure p_f, a volume V_f, and a temperature T_f. In this process energy may be transferred form or into the system and also work can be done by or on the system. One example of a thermodynamic process is increasing the pressure of a gas while maintaining a constant temperature.

Isobaric Process

An isobaric process is a thermodynamic process, in which the pressure of the system remains constant (p = const). The heat transfer into or out of the system does work, but also changes the internal energy of the system.

Since there are changes in internal energy (dU) and changes in system volume (ΔV), engineers often use the enthalpy of the system, which is defined as:

$$H = U + pV$$

In many thermodynamic analyses it is convenient to use the enthalpy instead of the internal energy. Especially in case of the first law of thermodynamics.

The enthalpy is the preferred expression of system energy that changes in many chemical, biological, and physical measurements at constant pressure. It is so useful that it is tabulated in the steam tables along with specific volume and specific internal energy. It is due to the fact, it simplifies the description of energy transfer. At constant pressure, the enthalpy change equals the energy transferred from the environment through heating ($Q = H_2 - H_1$) or work other than expansion work. For a variable-pressure process, the difference in enthalpy is not quite as obvious.

There are expressions in terms of more familiar variables such as temperature and pressure:

$$dH = C_p dT + V(1 - \alpha T)dp$$

Where C_p is the heat capacity at constant pressure and **α** is the coefficient of (cubic) thermal expansion. For ideal gas αT = 1 and therefore:

$$dH = C_p dT$$

For an ideal gas and a polytropic process, the case n = 0 corresponds to an isobaric (constant-pressure) process. In contrast to adiabatic process, in which n = *and a system exchanges no heat with its surroundings (Q = 0; ΔT ≠ 0)*, in an isobaric process there is a change in the internal energy (due to ΔT ≠ 0) and therefore ΔU ≠ 0 (for ideal gases) and Q ≠ 0.

In engineering, both very important thermodynamic cycles (Brayton and Rankine cycle) are based on two isobaric processes, therefore the study of this process is crucial for power plants.

First Law	$dU = dQ - pdV$ $dH = dQ - Vdp = dQ$
Ideal Gas Relation	$\frac{V}{T} = \text{constant}$
p, V, T Relations	$\frac{V_1}{T_1} = \frac{V_2}{T_2}$
Change in Internal Energy	$dU = mc_V(T_2 - T_1)$
Change in Enthalpy	$dH = mc_p(T_2 - T_1)$
Heat Transfer	$dQ = mc_p(T_2 - T_1)$
pdV Work	$W_{i \to f} = p(V_f - V_i)$ $W_{i \to f} = nR(T_f - T_i)$
Vdp Work	$W_{i \to f} = 0$

Isobaric Process and the First Law

The classical form of the first law of thermodynamics is the following equation:

$$dU = dQ - dW$$

In this equation dW is equal to dW = pdV and is known as the boundary work.

In an isobaric process and the ideal gas, part of heat added to the system will be used to do work and part of heat added will increase the internal energy (increase the temperature). Therefore it is convenient to use the enthalpy instead of the internal energy. Since H = U + pV, therefore dH = dU + pdV + Vdp and we substitute dU = dH – pdV – Vdp into the classical form of the law:

$$dH - pdV - Vdp = dQ - pdV$$

We obtain the law in terms of enthalpy:

$$dH = dQ + Vdp$$

or

$$dH = TdS + Vdp$$

In this equation the term Vdp is a flow process work. This work, Vdp, is used for open flow systems like a turbine or a pump in which there is a "dp", i.e. change in pressure. There are no changes in control volume. As can be seen, this form of the law simplifies the description of energy transfer. At constant pressure, the enthalpy change equals the energy transferred from the environment through heating:

Isobaric process (Vdp = 0):

$$dH = dQ \rightarrow Q = H_2 - H_1$$

At constant entropy, i.e. in isentropic process, the enthalpy change equals the flow process work done on or by the system.

Isentropic process (dQ = 0):

$$dH = Vdp \rightarrow W = H_2 - H_1$$

It is obvious, it will be very useful in analysis of both thermodynamic cycles used in power engineering, i.e. in Brayton cycle and Rankine cycle.

Example of Frictionless Piston – Heat – Enthalpy

A frictionless piston is used to provide a constant pressure of 500 kPa in a cylinder containing steam (superheated steam) of a volume of 2 m³ at 500 K. Calculate the final temperature, if 3000 kJ of heat is added.

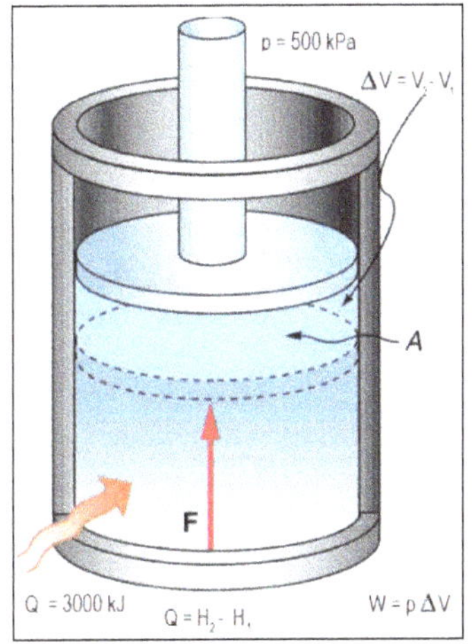

Calculate the final temperature, if 3000 kJ of heat is added.

Solution:

Using steam tables we know, that the specific enthalpy of such steam (500 kPa; 500 K) is about 2912 kJ/kg. Since at this condition the steam has density of 2.2 kg/m^3, then we know there is about 4.4 kg of steam in the piston at enthalpy of 2912 kJ/kg x 4.4 kg = 12812 kJ.

When we use simply $Q = H_2 - H_1$, then the resulting enthalpy of steam will be:

$$H_2 = H_1 + Q = 15812 \text{ kJ}$$

From steam tables, such superheated steam (15812/4.4 = 3593 kJ/kg) will have a temperature of 828 K (555 °C). Since at this enthalpy the steam have density of 1.31 kg/m^3, it is obvious that it has expanded by about 2.2/1.31 = 1.67 (+67%). Therefore the resulting volume is 2 m^3 x 1.67 = 3.34 m^3 and ΔV = 3.34 m^3 – 2 m^3 = 1.34 m^3.

The $p\Delta V$ part of enthalpy, i.e. the work done is:

$$W = p\Delta V = 500\,000 \text{ Pa} \times 1.34 \text{ m}^3 = 670 \text{ kJ}$$

Isobaric Process – Ideal Gas Equation

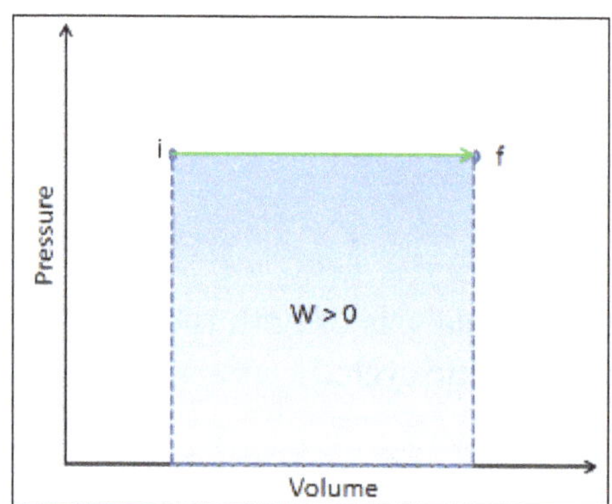

On a p-V diagram, the process occurs along a horizontal line (called an isobar) that has the equation p = constant.

Let assume an isobaric heat addition in an ideal gas. In an ideal gas, molecules have no volume and do not interact. According to the ideal gas law, pressure varies linearly with temperature and quantity, and inversely with volume.

$$pV = nRT$$

where:

- p is the absolute pressure of the gas.
- n is the amount of substance.
- T is the absolute temperature.
- V is the volume.
- R is the ideal, or universal, gas constant, equal to the product of the Boltzmann constant and the Avogadro constant.

In this equation the symbol R is a constant called the universal gas constant that has the same value for all gases—namely, R = 8.31 J/mol K.

The isobaric process can be expressed with the ideal gas law as:

$$\frac{V}{T} = \text{constant}$$

or

$$\frac{V_1}{T_1} = \frac{V_2}{T_2}$$

On a p-V diagram, the process occurs along a horizontal line (called an isobar) that has the equation p = constant.

Pressure-volume work by the closed system is defined as:

$$W = \int_{V_i}^{V_f} PdV$$

$$\delta W = PdV$$

Assuming that the quantity of ideal gas remains constant and applying the ideal gas law, this becomes.

$$W_{i \to f} = p\left(V_f - V_i\right) = nR\left(T_f - T_i\right)$$

According to the ideal gas model, the internal energy can be calculated by:

$$\Delta U = m\ c_v \Delta T$$

Where the property c_v *(J/*mol K) is referred to as specific heat *(or* heat capacity) at a constant volume because under certain special conditions (constant volume) it relates the temperature change of a system to the amount of energy added by heat transfer.

Adding these equations together, we obtain the equation for heat:

$$\mathrm{Q} = m\ c_v \Delta T + m\ R\Delta T = \ m\ (c_v + R)\Delta T = m c_p \Delta T$$

Where the property c_p *(J/*mol K) is referred to as specific heat *(or* heat capacity) at a constant pressure.

Charles's Law

Charles's Law is one of the gas laws. At the end of the 18th century, a French inventor and scientist Jacques Alexandre César Charles studied the relationship between the volume and the temperature of a gas at constant pressure. The results of certain experiments with gases at relatively low pressure led Jacques Alexandre César Charles to formulate a well-known law. It states that:

> For a fixed mass of gas at constant pressure, the volume is directly proportional to the Kelvin temperature.

That means that, for example, if you double the temperature, you will double the volume. If you halve the temperature, you will halve the volume.

You can express this mathematically as:

$$V = \text{constant} \,.\, T$$

Yes, it seems to be identical as isobaric process of ideal gas. These results are fully consistent with ideal gas law, which determinates, that the constant is equal to nR/p. If you rearrange the pV = nRT equation by dividing both sides by p, you will obtain:

$$V = nR/p \,.\, T$$

where nR/p is constant and:

- p is the absolute pressure of the gas.
- n is the amount of substance.
- T is the absolute temperature.
- V is the volume.
- R is the ideal, or universal, gas constant, equal to the product of the Boltzmann constant and the Avogadro constant.

In this equation the symbol R is a constant called the universal gas constant that has the same value for all gases—namely, R = 8.31 J/mol K.

Example of Isobaric Process – Isobaric Heat Addition

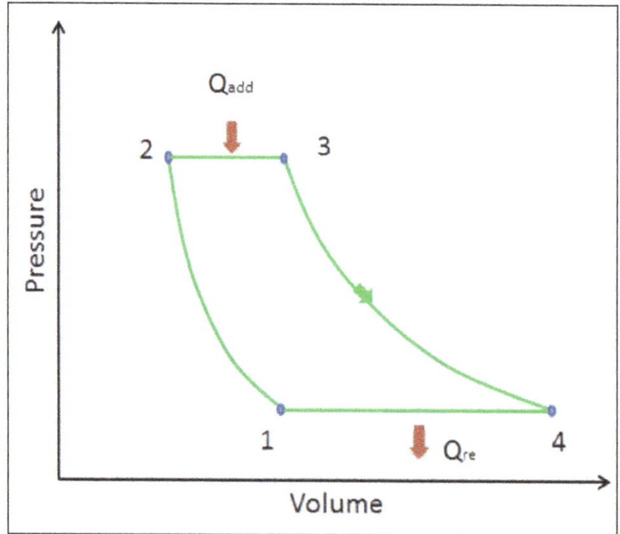

Ideal Brayton cycle consist of four thermodynamic processes. Two isentropic processes and two isobaric processes.

Let assume the ideal Brayton cycle that describes the workings of a constant pressure heat engine. Modern gas turbine engines and airbreathing jet engines also follow the Brayton cycle.

Ideal Brayton cycle consist of four thermodynamic processes. Two isentropic processes and two isobaric processes.

- Isentropic compression – Ambient air is drawn into the compressor, where it is pressurized ($1 \rightarrow 2$). The work required for the compressor is given by $W_C = H_2 - H_1$.
- Isobaric heat addition – The compressed air then runs through a combustion chamber, where fuel is burned and air or another medium is heated ($2 \rightarrow 3$). It is a constant-pressure process, since the chamber is open to flow in and out. The net heat added is given by $Q_{add} = H_3 - H_2$.
- Isentropic expansion – The heated, pressurized air then expands on turbine, gives up its energy. The work done by turbine is given by $W_T = H_4 - H_3$.
- Isobaric heat rejection – The residual heat must be rejected in order to close the cycle. The net heat rejected is given by $Q_{re} = H_4 - H_1$.

Assume an isobaric heat addition ($2 \rightarrow 3$) in a heat exchanger. In typical gas turbines the high-pressure stage receives gas (point 3 at the figure; $p_3 = 6.7$ MPa; $T_3 = 1190$ K (917 °C)) from a heat exchanger. Moreover we know, that the compressor receives gas (point 1 at the figure; $p_1 = 2.78$ MPa; $T_1 = 299$ K (26 °C)) and we know that the isentropic efficiency of the compressor is $\eta_K = 0.87$ (87%).

Calculate the Heat Added by the Heat Exchanger (between 2 → 3).

Solution:

From the first law of thermodynamics, the net heat added is given by $Q_{add} = H_3 - H_2$ or $Q_{add} = C_p.(T_3 - T_{2s})$, but in this case we do not know the temperature (T_{2s}) at the outlet of the compressor. We will solve this problem in intensive variables. We have to rewrite the previous equation (to include η_K) using the term $(+h_1 - h_1)$ to:

$$Q_{add} = h_3 - h_2 = h_3 - h_1 - (h_{2s} - h_1)/\eta_K$$

$$Q_{add} = c_p(T_3 - T_1) - (c_p(T_{2s} - T_1)/\eta_K)$$

Then we will calculate the temperature, T_{2s}, using p, V, T Relation for adiabatic process between ($1 \rightarrow 2$).

$$\frac{p_1}{p_2} = \left[\frac{V_2}{V_1}\right]^{\kappa} = \left[\frac{T_1}{T_2}\right]^{\frac{\kappa}{\kappa-1}}$$

In this equation the factor for helium is equal to $=c_p/c_v=1.66$. From the previous equation follows that the compressor outlet temperature, T_{2s}, is:

$$T_{2s} = \left[\frac{p_2}{p_1}\right]^{\frac{\kappa-1}{\kappa}} .T_1 = \left[\frac{6.7}{2.78}\right]^{\frac{0.66}{1.66}} .299 = 424\,K\,(151°C)$$

From Ideal Gas Law we know, that the molar specific heat of a monatomic ideal gas is:

$$C_v = 3/2R = 12.5\ \text{J/mol K and } C_p = C_v + R = 5/2R = 20.8\ \text{J/mol K}$$

We transfer the specific heat capacities into units of J/kg K via:

$$c_p = C_p \,.\, 1/M\ \text{(molar weight of helium)} = 20.8 \times 4.10^{-3} = 5200\ \text{J/kg K}$$

Using this temperature and the isentropic compressor efficiency we can calculate the heat added by the heat exchanger:

$$Q_{add} = c_p(T_3\text{-}T_1) - (c_p(T_{2s}\text{-}T_1)/\eta_K) = 5200.(1190 - 299) - 5200.(424\text{-}299)/0.87 = 4.633\ \text{MJ/kg} - 0.747\ \text{MJ/kg} = 3.886\ \text{MJ/kg}$$

Isobaric Process and Phase Diagrams

In a phase diagram, an isobaric process would show up as a horizontal line, since it takes place under a constant pressure. This diagram would show you at what temperatures a substance is solid, liquid, or vapor for a range of atmospheric pressures.

Isochoric Process

An isochoric process is a thermodynamic process, in which the volume of the closed system remains constant (V = const). It describes the behavior of gas inside the container, that cannot be deformed. Since the volume remains constant, the heat transfer into or out of the system does not the pΔV work, but only changes the internal energy (the temperature) of the system.

For an ideal gas and a polytropic process, the case $n \to \infty$ corresponds to an isochoric (constant-volume) process. In contrast to adiabatic process, in which n = and a system exchanges no heat with its surroundings ($Q = 0$; $W \neq 0$), in an isochoric process there is a change in the internal energy (due to $\Delta T \neq 0$) and therefore $\Delta U \neq 0$ (for ideal gases) and ($Q \neq 0$; $W = 0$).

In engineering of internal combustion engines, isochoric processes are very important for their thermodynamic cycles (Otto and Diesel cycle), therefore the study of this process is crucial for automotive engineering.

Isochoric Process and the First Law

The classical form of the first law of thermodynamics is the following equation:

$$dU = dQ - dW$$

In this equation dW is equal to dW = pdV and is known as the boundary work. Then:

$$dU = dQ - pdV$$

In isochoric process and the ideal gas, all of heat added to the system will be used to increase the internal energy.

Isochoric process (pdV = 0):

dU = dQ (for ideal gas)

First Law	dU = dQ
Ideal Gas Relation	$\frac{p}{T} = \text{constant}$
p, V, T Relations	$\frac{p_1}{T_1} = \frac{p_2}{T_2}$
Change in Internal Energy	$dU = mc_V\left(T_2 - T_1\right)$
Heat Transfer	$dQ = mc_V\left(T_2 - T_1\right)$
pdV Work	$W_{i \to f} = 0$

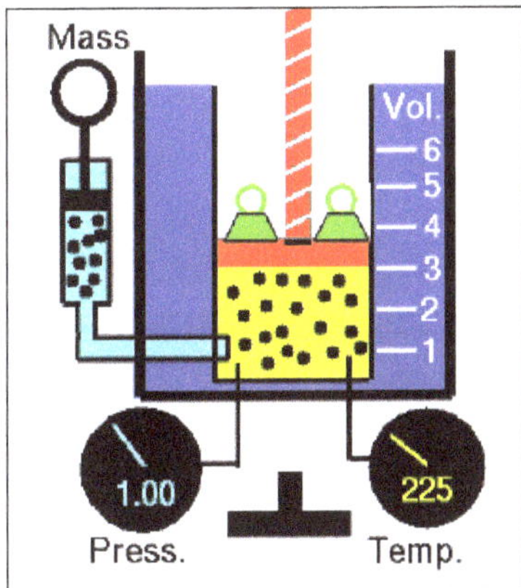

For a fixed mass of gas at constant volume, the pressure is directly proportional to the Kelvin temperature.

Isochoric Process – Ideal Gas Equation

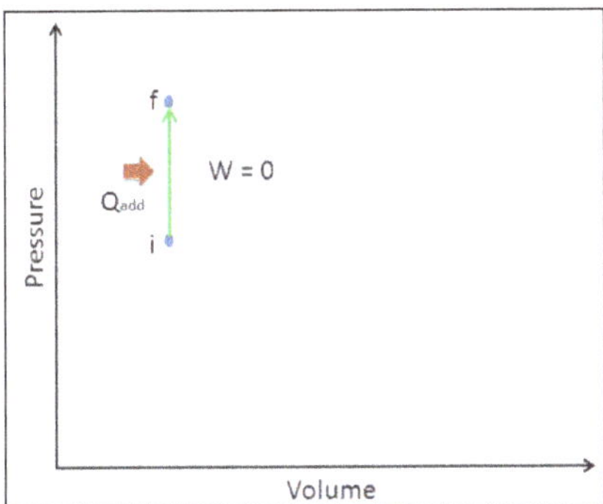

Let assume an isochoric heat addition in an ideal gas. In an ideal gas, molecules have no volume and do not interact. According to the ideal gas law, pressure varies linearly with temperature and quantity, and inversely with volume.

pV = nRT

where:

- p is the absolute pressure of the gas.
- n is the amount of substance.
- T is the absolute temperature.
- V is the volume.
- R is the ideal, or universal, gas constant, equal to the product of the Boltzmann constant and the Avogadro constant.

In this equation the symbol R is a constant called the universal gas constant that has the same value for all gases—namely, R = 8.31 J/mol K.

The isochoric process can be expressed with the ideal gas law as:

$$\frac{p}{T} = \text{constant}$$

or

$$\frac{p_1}{T_1} = \frac{p_2}{T_2}$$

On a p-V diagram, the process occurs along a horizontal line that has the equation V = constant.

Pressure-volume work by the closed system is defined as:

$$W = \int_{V_i}^{V_f} P\,dV$$

$$\delta W = PdV$$

Since the process is isochoric, dV = 0, the pressure-volume work is equal to zero. According to the ideal gas model, the internal energy can be calculated by:

$$\Delta U = m\, c_v \Delta T$$

where the property c_v *(J/*mol K) is referred to as specific heat *(or* heat capacity) at a constant volume because under certain special conditions (constant volume) it relates the temperature change of a system to the amount of energy added by heat transfer.

Since there is no work done by or on the system, the first law of thermodynamics dictates ΔU = ΔQ. Therefore:

$$Q = m\, c_v \Delta T$$

Guy-Lussac's Law

Guy-Lussac's Law or the Pressure Law, one of the gas laws, states that:

For a fixed mass of gas at constant volume, the pressure is directly proportional to the Kelvin temperature.

That means that, for example, if you double the temperature, you will double the pressure. If you halve the temperature, you will halve the pressure.

You can express this mathematically as:

$$p = \text{constant} \, . \, T$$

Yes, it seems to be identical as isochoric process of ideal gas. These results are fully consistent with ideal gas law, which determinates, that the constant is equal to nR/V. If you rearrange the pV = nRT equation by dividing both sides by V, you will obtain:

$$p = nR/V \, . \, T$$

where nR/V is constant and:

- p is the absolute pressure of the gas.
- n is the amount of substance.
- T is the absolute temperature.
- V is the volume.
- R is the ideal, or universal, gas constant, equal to the product of the Boltzmann constant and the Avogadro constant.

Example of Isochoric Process: Isochoric Heat Addition

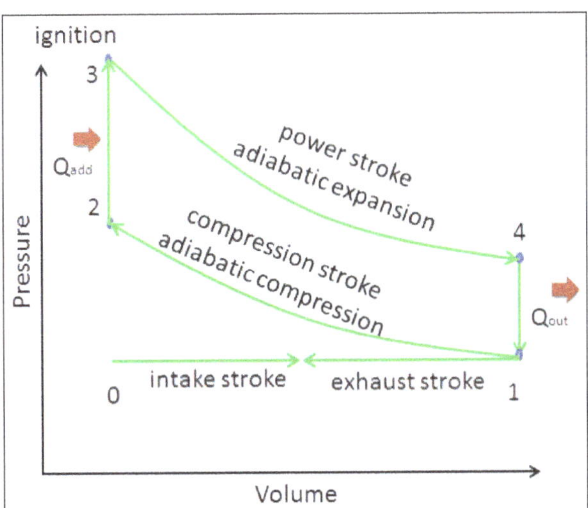

Let assume the Otto Cycle, which is the one of most common thermodynamic cycles *that can be found in* automobile engines. This cycle assumes that the heat addition occurs instantaneously (between 2 → 3) while the piston is at top dead center. This process is considered to be isochoric.

Processes 2 → 3 and 4 → 1 are isochoric processes, in which the heat is transferred into the system between 2 → 3 and out of the system between 4 → 1. During these processes no work is done on the system or extracted from the system. The isochoric process 2 → 3 is intended to represent the ignition of the fuel–air mixture and the subsequent rapid burning.

Example Calculation

Assume an isochoric heat addition in an ideal gas. In an ideal gas, molecules have no volume and do not interact. According to the ideal gas law, pressure varies linearly with temperature and quantity, and inversely with volume. The basic formula would be:

$$pV = nRT$$

where:

- p is the absolute pressure of the gas.
- n is the amount of substance.
- T is the absolute temperature.
- V is the volume.
- R is the ideal, or universal, gas constant equal to the product of the Boltzmann constant and the Avogadro constant.
- K is the scientific abbreviation for Kelvin.

In this equation the symbol R is a constant called the universal gas constant that has the same value for all gases—namely, R = 8.31 Joule/mole K.

The isochoric process can be expressed with the ideal gas law as:

$$p/T = \text{constant}$$

Since the process is isochoric, dV = 0, the pressure-volume work is equal to zero. According to the ideal gas model, the internal energy can be calculated by:

$$\Delta U = m\, c_v \Delta T$$

where the property c_v (J/mole K) is referred to as specific heat (or heat capacity) at a constant volume because under certain special conditions (constant volume) it relates the temperature change of a system to the amount of energy added by heat transfer.

Since there is no work done by or on the system, the first law of thermodynamics dictates ΔU = ΔQ. Therefore:

$$Q = m\, c_v \Delta T$$

Isothermal Process

An isothermal process is a thermodynamic process, in which the temperature of the system remains constant (T = const). The heat transfer into or out of the system typically must happen at such a slow rate in order to continually adjust to the temperature of the reservoir through heat exchange. In each of these states the thermal equilibrium is maintained.

For an ideal gas and a polytropic process, the case n = 1 corresponds to an isothermal (constant-temperature) process. In contrast to adiabatic process, in which n = κ and a system exchanges no heat with its surroundings (Q = 0; ΔT ≠ 0), in an isothermal process there is no change in the internal energy (due to ΔT = 0) and therefore ΔU = 0 (for ideal gases) and Q ≠ 0. An adiabatic process is not necessarily an isothermal process, nor is an isothermal process necessarily adiabatic.

In engineering, phase changes, such as evaporation or melting, are isothermal processes when, as is usually the case, they occur at constant pressure and temperature.

Isothermal Process and the First Law

The classical form of the first law of thermodynamics is the following equation:

$$dU = dQ - dW$$

In this equation dW is equal to dW = pdV and is known as the boundary work.

In isothermal process and the ideal gas, all heat added to the system will be used to do work:

Isothermal process (dU = 0):

$$dU = 0 = Q - W \rightarrow W = Q \quad \text{(for ideal gas)}$$

First Law	$dU = 0$ $dQ = dW$
Ideal Gas Relation	pV = constant
p, V, T Relations	$p_1V_1 = p_2V_2$
Change in Internal Energy	dU = 0
Change in Enthalpy	dH = 0
Heat Transfer	$Q = nRT \ \text{In}\frac{V_f}{V_i}$ $Q = p_iV_i \ \text{In}\frac{V_f}{V_i}$
pdV Work	$W_{i \rightarrow f} = nRT \ \text{In}\frac{V_f}{V_i}$ $W_{i \rightarrow f} = p_iV_i \ \text{In}\frac{V_f}{V_i}$

Isothermal Expansion: Isothermal Compression

In an ideal gas, molecules have no volume and do not interact. According to the ideal gas law, pressure varies linearly with temperature and quantity, and inversely with volume:

$$pV = nRT$$

where:

- p is the absolute pressure of the gas.
- n is the amount of substance.
- T is the absolute temperature.
- V is the volume.
- R is the ideal, or universal, gas constant, equal to the product of the Boltzmann constant and the Avogadro constant.

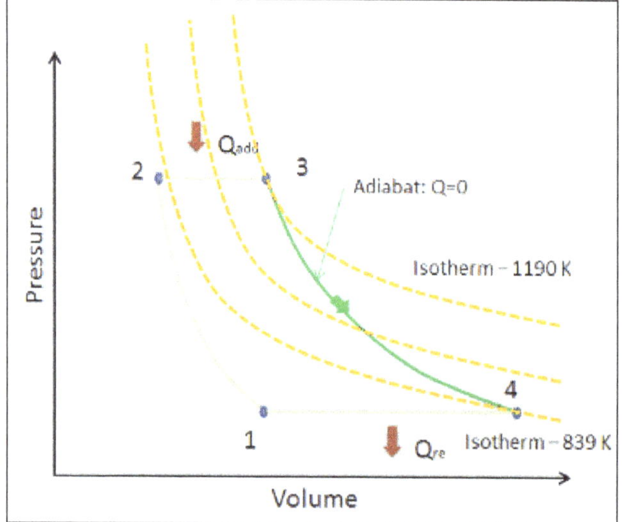

In this equation the symbol R is a constant called the universal gas constant that has the same value for all gases—namely, R = 8.31 J/mol K.

The isothermal process can be expressed with the ideal gas law as:

$$pV = \text{constant}$$

Or

$$p_1V_1 = p_2V_2$$

On a p-V diagram, the process occurs along a line (called an isotherm) that has the equation p = constant/V.

Boyle–Mariotte Law

Boyle-Mariotte Law is one of the gas laws. At the end of the 17th century, Robert William Boyle and Edme Mariotte independently studied the relationship between the volume and pressure of a gas at constant temperature. The results of certain experiments with gases at relatively low pressure led Robert Boyle to formulate a well-known law. It states that:

For a fixed mass of gas at constant temperature, the volume is inversely proportional to the pressure.

That means that, for example, if you increase the volume 10 times, the pressure will decrease 10 times. If you halve the volume, you will double the pressure.

You can express this mathematically as:

$$pV = \text{constant}$$

Or

$$p_1V_1 = p_2V_2$$

Yes, it seems to be identical as isothermal process of ideal gas. In fact, during their experiments the temperature remain constant as was assumed by Mariotte. These results are fully consistent with ideal gas law, which determinates, that the constant is equal to nRT:

$$pV = nRT$$

where:

- *p* is the absolute pressure of the gas.
- *n* is the amount of substance.
- *T* is the absolute temperature.
- *V* is the volume.
- *R* is the ideal, or universal, gas constant, equal to the product of the Boltzmann constant and the Avogadro constant.

In this equation the symbol R is a constant called the universal gas constant that has the same value for all gases—namely, R = 8.31 J/mol K.

Example of Isothermal Process

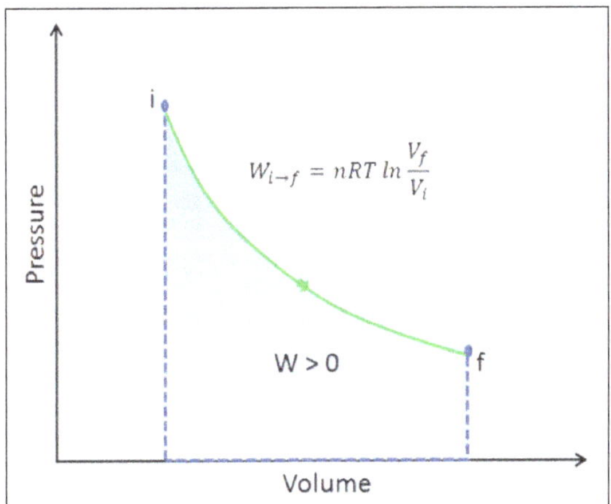

Isothermal process – pV diagram.

Assume an isothermal expansion of helium (i → f) in a frictionless piston (closed system). The gas expansion is propelled by absorption of heat energy Q_{add}. The gas expands from initial volume of 0.001 m³ and simultaneously the external load of the piston slowly and continuously decreases from 1 MPa to 0.5 MPa. Since helium behaves almost as an ideal gas, use the ideal gas law to calculate final volume of the chamber and then calculate the work done by the system, when the temperature of the gas is equal to 400 K.

Solution:

The final volume of the gas, V_f, can be calculated using p, V, T Relation for isothermal process:

$$p_i V_i = p_f V_f \Rightarrow V_f = p_i V_i / p_f = 2 \text{ x } 0.001 \text{ m}^3 = 0.002 \text{ m}^3$$

To calculate such processes, we would need to know how pressure varies with volume for the actual process by which the system changes from state i to state f. Since during this process the internal pressure was not constant, the pΔV work done by the piston must be calculated using the following integral:

$$W_{i \to f} = \int_{V_i}^{V_f} p \; dV = \int_{V_i}^{V_f} \frac{nRT}{V} dV = nRT \int_{V_i}^{V_f} \frac{1}{V} dV = nRT \; In \frac{V_f}{V_i}$$

$$W_{i \to f} = const. \; In \frac{V_f}{V_i} = 10^6 \left[MPa \right] .0.001 \left[m^3 \right] . \; In2 = 693 \; J$$

By convention, a positive value for work indicates that work is done by the system on its surroundings. A negative value indicates that work is done on the system by its surroundings. The pΔV work is equal to the area under the process curve plotted on the pressure-volume diagram.

Free Expansion – Joule Expansion

These are adiabatic processes in which no transfer of heat occurs between the system and its environment and no work is done on or by the system. These types of adiabatic processes are called free expansion. It is an irreversible process in which a gas expands into an insulated evacuated chamber. It is also called Joule expansion. For an ideal gas, the temperature doesn't change (this means that the process is also isothermal), however, real gases experience a temperature change during free expansion. In free expansion Q = W = 0, and the first law requires that:

$$dE_{int} = 0$$

A free expansion can not be plotted on a P-V diagram, because the process is rapid, not quasistatic. The intermediate states are not equilibrium states, and hence the pressure is not clearly defined.

Adiabatic Process

An adiabatic process is a thermodynamic process, in which there is no heat transfer into or out of the system (Q = 0). The system can be considered to be perfectly insulated. In an adiabatic process, energy is transferred only as work. The assumption of no heat transfer is very important, since we can use the adiabatic approximation only in very rapid processes. In these rapid processes, there is not enough time for the transfer of energy as heat to take place to or from the system.

In real devices (such as turbines, pumps, and compressors) heat losses and losses in the combustion process occur, but these losses are usually low in comparison to overall energy flow and we can approximate some thermodynamic processes by the adiabatic process.

Adiabitic vs. Isentropic Process

In comparison to the isentropic process in which the entropy of the fluid or gas remains constant, in the adiabatic process the entropy changes. Therefore the adiabatic process is considered to be irreversible process. The isentropic process is a special case of an adiabatic process. The isentropic process is a reversible adiabatic process. An isentropic process can also be called a constant entropy process. In engineering such an idealized process is very useful for comparison with real processes.

One way to make real processes approximate reversible process is to carry out the process in a series of small or infinitesimal steps or infinitely slowly, so that the process can be considered as a series of equilibrium states. For example, heat transfer may be considered reversible if it occurs due to a small temperature difference between the system and its surroundings. But real processes are not done infinitely slowly. Reversible processes are a useful and convenient theoretical fiction, but do not occur in nature. For example, there could be turbulence in the gas. Therefore, heat engines must have lower efficiencies than limits on their efficiency due to the inherent irreversibility of the heat engine cycle they use.

Adiabatic Process and the First Law

For a closed system, we can write the first law of thermodynamics in terms of enthalpy:

$$dH = dQ + Vdp$$

In this equation the term Vdp is a flow process work. This work, Vdp, is used for open flow systems like a turbine or a pump in which there is a "dp", i.e. change in pressure. As can be seen, this form of the law simplifies the description of energy transfer. In adiabatic process, the enthalpy change equals the flow process work done on or by the system:

First Law	dU = pdV dH = Vdp
Ideal Gas Relation	$pV^k = \text{constant}$
p, V, T Relations	$\frac{p_1}{p_2} = \left[\frac{V_2}{V_1}\right]^{\kappa} = \left[\frac{T_1}{T_2}\right]^{\frac{\kappa}{\kappa-1}}$
Change in Internal Energy	$dU = mc_V\left(T_2 - T_1\right)$
Change in Enthalpy	$dH = mc_p\left(T_2 - T_1\right)$
Heat Transfer	0

pdV Work	$W_{dV} = pdV = dU = mc_V(T_2 - T_1)$
Vdp Work	$W_{dp} = Vdp = dH = mc_p(T_2 - T_1)$

Adiabatic process (dQ = 0):

$$dH = Vdp \rightarrow W = H_2 - H_1 \rightarrow H_2 - H_1 = C_p(T_2 - T_1) \text{ (for an ideal gas)}$$

Adiabatic Expansion: Adiabatic Compression

In an ideal gas, molecules have no volume and do not interact. According to the ideal gas law, pressure varies linearly with temperature and quantity, and inversely with volume:

$$pV = nRT$$

where:

- p is the absolute pressure of the gas.
- n is the amount of substance.
- T is the absolute temperature.
- V is the volume.
- R is the ideal, or universal, gas constant, equal to the product of the Boltzmann constant and the Avogadro constant.

In this equation the symbol R is a constant called the universal gas constant that has the same value for all gases—namely, R = 8.31 J/mol K.

The adiabatic process can be expressed with the ideal gas law as:

$$pV^{\kappa} = constant$$

Or

$$p_1 V_1^{\kappa} = p_2 V_2^{\kappa}$$

In which $\kappa = c_p/c_v$ is the ratio of the specific heats (or heat capacities) for the gas. One for constant pressure (c_p) and one for constant volume (c_v). Note that, this ratio $\kappa = c_p/c_v$ is a factor in determining the speed of sound in a gas and other adiabatic processes.

Other p, V and T Relation

$$\frac{p_1}{p_2} = \left[\frac{V_2}{V_1}\right]^{\kappa} = \left[\frac{T_1}{T_2}\right]^{\frac{\kappa}{\kappa-1}}$$

On a p-V diagram, the process occurs along a line (called an adiabat) that has the equation p = constant / V^κ. For an ideal gas and a polytropic process, the case $n = \kappa$ corresponds to an adiabatic process.

Example of Adiabatic Expansion

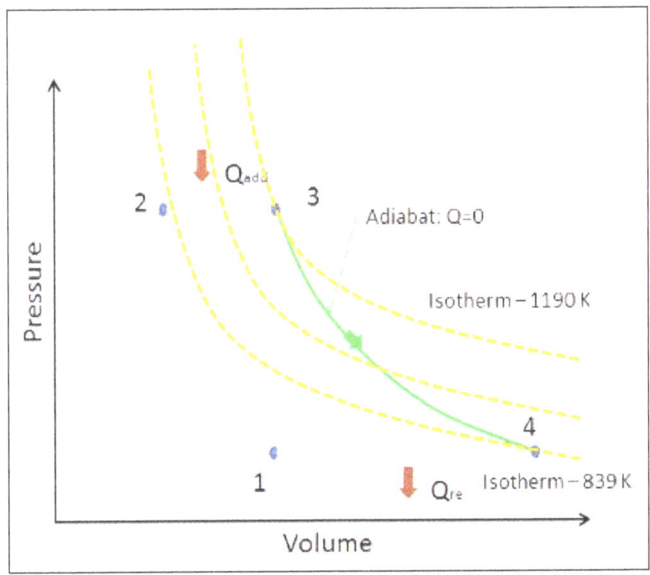

Assume an adiabatic expansion of helium (3 → 4) in a gas turbine (Brayton cycle).

Assume an adiabatic expansion of helium (3 → 4) in a gas turbine. Since helium behaves almost as an ideal gas, use the ideal gas law to calculate outlet temperature of the gas ($T_{4,real}$). In this turbines the high-pressure stage receives gas (point 3 at the figure; p_3 = 6.7 MPa; T_3 = 1190 K (917°C)) from a heat exchanger and exhaust it to another heat exchanger, where the outlet pressure is p_4 = 2.78 MPa (point 4).

Solution:

The outlet temperature of the gas, $T_{4,real}$, can be calculated using p, V, T Relation for adiabatic process. Note that, it is the same relation as for the isentropic process, therefore results must be identical. It this case, we calculate the expansion for different gas turbine (less efficient) as in case of Isentropic Expansion in Gas Turbine.

$$\frac{p_1}{p_2} = \left[\frac{V_2}{V_1}\right]^\kappa = \left[\frac{T_1}{T_2}\right]^{\frac{\kappa}{\kappa-1}}$$

In this equation the factor for helium is equal to $\kappa = c_p/c_v$=1.66. From the previous equation follows that the outlet temperature of the gas, $T_{4,real}$, is:

$$T_{4,real} = \left[\frac{p_4}{p_3}\right]^{\frac{\kappa-1}{\kappa}} .T_3 = \left[\frac{2.78}{6.7}\right]^{\frac{0.66}{1.66}} .1190 = 839\,K\ (566\ °C)$$

Adiabatic Process in Gas Turbine

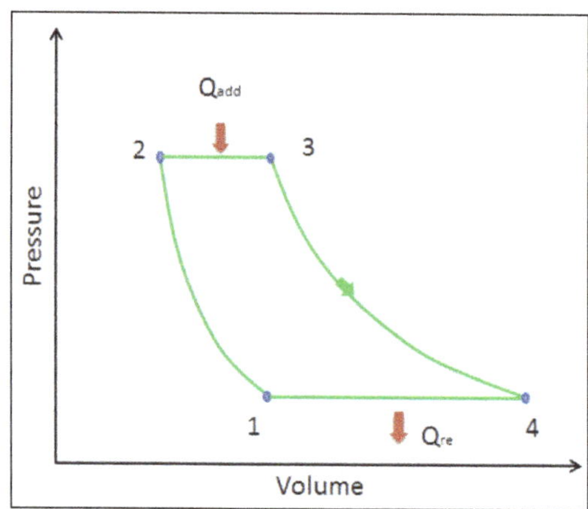

Ideal Brayton cycle consist of four thermodynamic processes.
Two isentropic processes and two isobaric processes.

Let assume the Brayton cycle that describes the workings of a constant pressure heat engine. Modern gas turbine engines and airbreathing jet engines also follow the Brayton cycle.

The Brayton cycle consist of four thermodynamic processes. Two adiabatic processes and two isobaric processes.

- Adiabatic compression – Ambient air is drawn into the compressor, where it is pressurized ($1 \rightarrow 2$). The work required for the compressor is given by $W_C = H_2 - H_1$.
- Isobaric heat addition – The compressed air then runs through a combustion chamber, where fuel is burned and air or another medium is heated ($2 \rightarrow 3$). It is a constant-pressure process, since the chamber is open to flow in and out. The net heat added is given by $Qadd = H_3 - H_2$.
- Adiabatic expansion – The heated, pressurized air then expands on turbine, gives up its energy. The work done by turbine is given by $W_T = H_4 - H_3$.
- Isobaric heat rejection – The residual heat must be rejected in order to close the cycle. The net heat rejected is given by $Q_{re} = H_4 - H_1$.

As can be seen, we can describe and calculate (e.g. thermal efficiency) such cycles (similarly for Rankine cycle) using enthalpies.

Isentropic Efficiency: Turbine, Compressor and Nozzle

We assumed that the gas expansion is isentropic and therefore we used $T_{4,is}$ as the outlet temperature of the gas. These assumptions are only applicable with ideal cycles.

Most steady-flow devices (turbines, compressors, nozzles) operate under adiabatic conditions, but they are not truly isentropic but are rather idealized as isentropic for calculation purposes. We define parameters η_T, η_C, η_N, as a ratio of real work done by device to work by device when operated under isentropic conditions (in case of turbine). This ratio is known as the Isentropic Turbine/Compressor/Nozzle Efficiency.

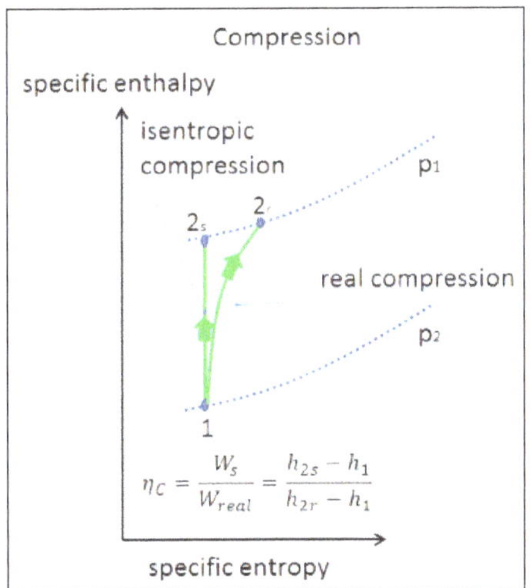

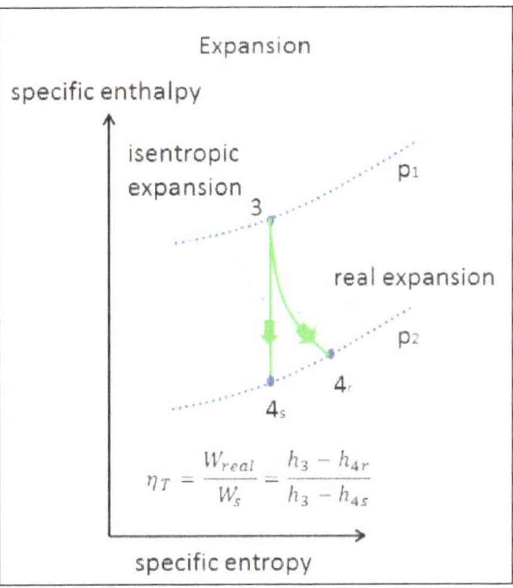

Isentropic process is a special case of adiabatic processes. It is a reversible adiabatic process. An isentropic process can also be called a constant entropy process.

These parameters describe how efficiently a turbine, compressor or nozzle approximates a corresponding isentropic device. This parameter reduces the overall efficiency and work output. For turbines, the value of $\boldsymbol{\eta}_T$ is typically 0.7 to 0.9 (70–90%).

Isentropic Turbine Efficiency:

$$\eta_T = \frac{\text{Real Turbine Work}}{\text{Isentropic Turbine Work}} = \frac{Wreal}{W_S} = \frac{h_1 - h_{2r}}{h_1 - h_{2s}}$$

Isentropic Compressor Efficiency:

$$\eta_C = \frac{\text{Isentropic Compressor work}}{\text{Real Compressor Work}} = \frac{W_s}{Wreal} = \frac{h_{2s} - h_1}{h_{2r} - h_1}$$

where:

- h_1 is the specific enthalpy of the gas at the entrance.
- h_{2r} is the specific enthalpy of the gas at the exit for real process.
- h_{2s} is the specific enthalpy of the gas at the exit for isentropic process.

Example of Isentropic Turbine Efficiency

Assume an isentropic expansion of helium ($3 \rightarrow 4$) in a gas turbine. In this turbines the high-pressure stage receives gas (point 3 at the figure; p_3 = 6.7 MPa; T_3 = 1190 K (917 °C)) from a heat exchanger and exhaust it to another heat exchanger, where the outlet pressure is p_4 = 2.78 MPa (point 4). The temperature (for isentropic process) of the gas at the exit of the turbine is T_{4s} = 839 K (566 °C).

Calculate the work done by this turbine and calculate the real temperature at the exit of the turbine, when the isentropic turbine efficiency is η_T = 0.91 (91%).

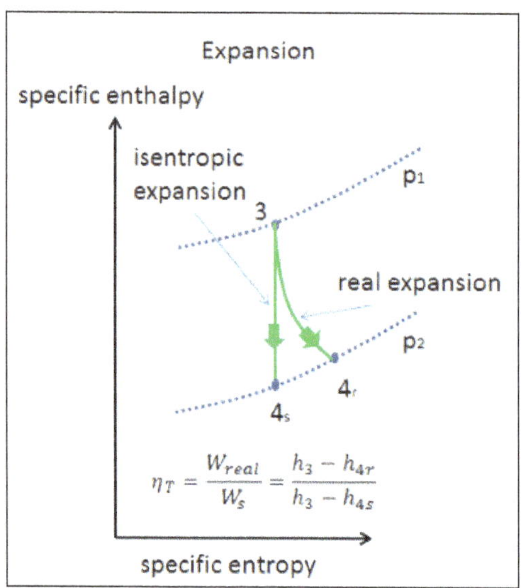

Isentropic process is a special case of adiabatic processes. It is a reversible adiabatic process. An isentropic process can also be called a constant entropy process.

Solution:

From the first law of thermodynamics, the work done by turbine in an isentropic process can be calculated from:

$$W_T = h_3 - h_{4s} \rightarrow W_{Ts} = c_p(T_3 - T_{4s})$$

From Ideal Gas Law we know, that the molar specific heat of a monatomic ideal gas is:

$$C_v = 3/2R = 12.5\ J/mol\ K \text{ and } C_p = C_v + R = 5/2R = 20.8\ J/mol\ K$$

We transfer the specific heat capacities into units of J/kg K via:

$$c_p = C_p.\ 1/M\left(\text{molar weight of helium}\right) = 20.8\ x\ 4.10^{-3} = \ 5200\ J/kg\ K$$

The work done by gas turbine in isentropic process is then:

$$W_{T,s} = c_p(T_3 - T_{4s}) = 5200\ x\left(1190 - 839\right) = 1.825\ MJ/kg$$

The real work done by gas turbine in adiabatic process is then:

$$W_{T,real} = c_p\ (T_3 - T_{4s})\ .\eta_T = 5200\ x\ \left(1190 - \ 839\right)\ x\ 0.91 = 1.661\ MJ/kg$$

Free Expansion: Joule Expansion

These are adiabatic processes in which no transfer of heat occurs between the system and its environment and no work is done on or by the system. These types of adiabatic processes are called free expansion. It is an irreversible process in which a gas expands into an insulated evacuated chamber. It is also called Joule expansion. For an ideal gas, the temperature doesn't change, however, real gases experience a temperature change during free expansion. In free expansion Q = W = 0, and the first law requires that:

$$d E_{int} = 0$$

A free expansion cannot be plotted on a P-V diagram, because the process is rapid, not quasistatic. The intermediate states are not equilibrium states, and hence the pressure is not clearly defined.

Time Scales and the Adiabatic Process

Although the theory of adiabatic process holds up when observed over long periods of time, smaller time scales render adiabatic impossible in mechanical processes—since there are no perfect insulators for isolated systems, heat is always lost when work is done.

In general, adiabatic processes are assumed to be those where the net outcome of temperature remains unaffected, though that does not necessarily mean that heat is not transferred throughout the process. Smaller time scales can reveal the minute transfer of heat over the system boundaries, which ultimately balance out over the course of work.

Factors such as the process of interest, the rate of heat dissipation, how much work is down, and the amount of heat lost through imperfect insulation can affect the outcome of heat transfer in the overall process, and for this reason, the assumption that a process is adiabatic relies on the observation of the heat transfer process as a whole instead of its smaller parts.

References

- Thermodynamic-process-and-their-types, thermal-physics: askiitians.com, Retrieved 14 March, 2019
- Thermodynamic-processes: reactor-physics.com, Retrieved 15 February, 2019
- Isobaric-process, thermodynamic-processes: nuclear-power.net, Retrieved 26 January, 2019
- Isobaric-process-2698984: nuclear-power.net, Retrieved 27 August, 2019
- Isochoric-process-2698985: thoughtco.com, Retrieved 25 May, 2019
- Isothermal-process, thermodynamic-processes: nuclear-power.net, Retrieved 07 June, 2019

4

Entropy

Entropy is the thermodynamic quantity which represents the degree of randomness in a thermodynamic system. Clausius Theorem, Clausius Inequality, Temperature-Entropy Diagrams, phase transition, etc. are some of the concepts that fall under its domain. This chapter discusses in detail all these concepts related to entropy.

Entropy is a function of the state of a thermodynamic system. It is a size-extensive quantity, invariably denoted by S, with dimension energy divided by absolute temperature (SI unit: joule/K). Entropy has no analogous mechanical meaning—unlike volume, a similar size-extensive state parameter. Moreover entropy cannot be measured directly, there is no such thing as an entropy meter, whereas state parameters like volume and temperature are easily determined. Consequently entropy is one of the least understood concepts in physics.

The state variable "entropy" was introduced by Rudolf Clausius in 1865, when he gave a mathematical formulation of the second law of thermodynamics.

The traditional way of introducing entropy is by means of a Carnot engine, an abstract engine conceived of by Sadi Carnot in 1824 as an idealization of a steam engine. Carnot's work foreshadowed the second law of thermodynamics. In this approach, entropy is the amount of heat (per degree kelvin) gained or lost by a thermodynamic system that makes a transition from one state to another. The second law states that the entropy of an isolated system increases in spontaneous (natural) processes leading from one state to another, whereas the first law states that the internal energy of the system is conserved.

In 1877 Ludwig Boltzmann gave a definition of entropy in the context of the kinetic gas theory, a branch of physics that developed into statistical thermodynamics. Boltzmann's definition of entropy was furthered by John von Neumann to a quantum statistical definition. In the statistical approach the entropy of an isolated (constant energy) system is $k_B \log\Omega$, where k_B is Boltzmann's constant and the function log stands for the natural (base e) logarithm. Ω is the number of different wave functions ("microstates") of the system belonging to the system's "macrostate" (thermodynamic state). The number Ω is the multiplicity of the macrostate; for an isolated system, where the macrostate is of definite energy, Ω is its degeneracy. For a system of about 10^{23} particles, Ω is on the order of $10^{10^{23}}$, that is the entropy is on the order of $10^{23} \times k_B \approx R$, the molar gas constant.

Not satisfied with the engineering type of argument, the mathematician Constantin Carathéodory gave in 1909 a new axiomatic formulation of entropy and the second law of thermodynamics. His theory was based on Pfaffian differential equations. His axiom replaced the earlier Kelvin-Planck and the equivalent Clausius formulation of the second law and did not need Carnot engines. Carathéodory's work was taken up by Max Born, and it is treated in a few monographs. Since it requires more mathematical knowledge than the traditional approach based on Carnot engines, the traditional approach, which depends on some ingenious thought experiments, is still dominant in the majority of introductory works on thermodynamics.

Specific Entropy

Entropy has corresponding intensive (size-independent) properties for pure materials. A corresponding intensive property is specific entropy, which is entropy per mass of substance involved. Specific entropy is denoted by a lower case *s*, with dimension of energy per absolute temperature and mass [SI unit: joule/(K·kg)]. If a molecular mass or number of moles involved can be assigned, then another corresponding intensive property is molar entropy, which is entropy per mole of the compound involved, or alternatively specific entropy times molecular mass. There is no universally agreed upon symbol for molar properties, and molar entropy has been at times confusingly symbolized by *S*, as in extensive entropy. The dimensions of molar entropy are energy per absolute temperature and number of moles [SI unit: joule/(K·mole)].

Traditional Definition of Entropy

The state (a point in state space) of a thermodynamic system is characterized by a number of variables, such as pressure p, temperature T, amount of substance n, volume V, etc. Any thermodynamic parameter can be seen as a function of an arbitrary independent set of other thermodynamic variables, hence the terms "property", "parameter", "variable" and "function" are used interchangeably. The number of independent thermodynamic variables of a system is equal to the number of energy contacts of the system with its surroundings.

An example of a reversible (quasi-static) energy contact is offered by the prototype thermodynamical system, a gas-filled cylinder with piston. Such a cylinder can perform work on its surroundings,

$$DW = pdV, \quad dV > 0,$$

Where dV stands for a small increment of the volume V of the cylinder, p is the pressure inside the cylinder and DW stands for a small amount of work, not necessarily a differential of a function; such differential is often referred to as inexact and indicated by a capital D, instead of d. Work by expansion is a form of energy contact between the cylinder and its surroundings. This process can be reverted, the volume of the cylinder can be decreased, the gas is compressed and the surroundings perform work DW = pdV < 0 on the cylinder.

When the inexact differential DW is divided by p, the quantity DW/p becomes obviously equal to the differential dV of the differentiable state function V. State functions depend only on the actual values of the thermodynamic parameters (they depend on a single point in state space, a state function is local in state space). A state function does not depend on the points on the path along which the state was reached (the history of the state). Mathematically this means that integration from point 1 to point 2 along path I in state space is equal to integration along a different path II,

$$V_2 - V_1 = \int_{1(I)}^{2} dV = \int_{1(II)}^{2} dV \Rightarrow \int_{1(I)}^{2} \frac{DW}{p} = \int_{1(II)}^{2} \frac{DW}{p}$$

The amount of work (divided by p) performed reversibly along path I is equal to the amount of work (divided by p) along path II. This condition is necessary and sufficient that DW/p is the differential of a state function. So, although DW is not a differential, the quotient DW/p is one.

Reversible absorption of a small amount of heat DQ is another energy contact of a system with its surroundings; DQ is again not a differential of a certain function. In a completely analogous manner to DW/p, the following result can be shown for the heat DQ (divided by T) absorbed reversibly by the system along two different paths (along both paths the absorption is reversible):

$$\int_{1(I)}^{2} \frac{DQ}{T} = \int_{1(II)}^{2} \frac{DQ}{T}.$$

Hence the quantity dS defined by:

$$dS \overset{\text{def}}{=} \frac{DQ}{T}$$

is the differential of a state variable S, the entropy of the system. In the next subsection equation (1) will be proved from the Kelvin-Planck principle. Observe that this definition of entropy only fixes entropy differences:

$$S_2 - S_1 \equiv \int_1^2 dS = \int_1^2 \frac{DQ}{T}$$

Note further that entropy has the dimension energy per degree temperature (joule per degree kelvin) and recalling the first law of thermodynamics (the differential dU of the internal energy satisfies dU = DQ – DW), it follows that,

$$dU = TdS - pdV.$$

(For convenience sake only a single work term was considered here, namely $DW = pdV$, work done *by* the system). The internal energy is an extensive quantity. The temperature T is an intensive property, independent of the size of the system. It follows that the entropy S is an extensive property. In that sense the entropy resembles the volume of the system. We reiterate that volume is a state function with a well-defined mechanical meaning, whereas entropy is introduced by analogy and is not easily visualized. Indeed, it requires a fairly elaborate reasoning to prove that S is a state function, i.e., that equation $\int_{1(I)}^{2} \frac{DQ}{T} = \int_{1(II)}^{2} \frac{DQ}{T}$ holds.

Proof that Entropy is a State Function

Equation $\int_{1(I)}^{2} \frac{DQ}{T} = \int_{1(II)}^{2} \frac{DQ}{T}$ gives the sufficient condition that the entropy S is a state function. The standard proof of equation:

$$\int_{1(I)}^{2} \frac{DQ}{T} = \int_{1(II)}^{2} \frac{DQ}{T}$$

as given now, is physical, by means of an engine making Carnot cycles, and is based on the Kelvin-Planck formulation of the second law of thermodynamics.

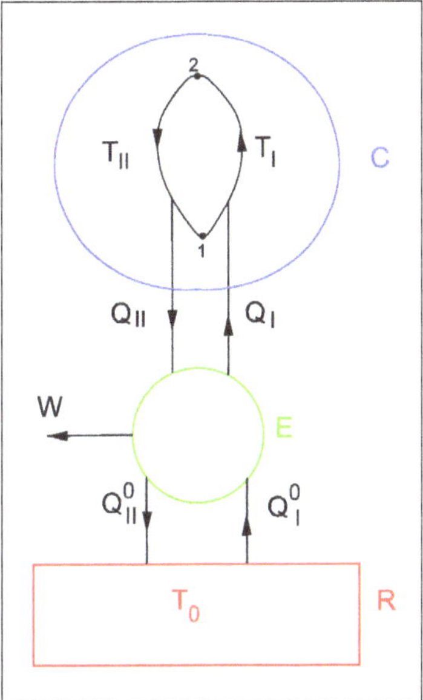

Consider the figure. A system, consisting of an arbitrary closed system *C* (only heat goes in and out) and a reversible heat engine *E*, is coupled to a large heat reservoir *R* of constant temperature T_0. The system *C* undergoes a cyclic state change 1-2-1. Since no work is performed on or by *C*, it follows that,

$$Q_{\text{I}} = Q_{\text{II}} \quad \text{with} \quad Q_{\text{I}} \equiv \int_1^2 DQ_{\text{I}}, \quad Q_{\text{II}} \equiv \int_1^2 DQ_{\text{II}}.$$

For the heat engine E it holds (by the definition of thermodynamic temperature) that:

$$\frac{DQ_{\text{I}}}{DQ_{\text{I}}^0} = \frac{T_{\text{I}}}{T_0} \quad \text{and} \quad \frac{DQ_{\text{II}}}{DQ_{\text{II}}^0} = \frac{T_{\text{II}}}{T_0}.$$

Hence,

$$\frac{Q_{\text{I}}^0}{T_0} \equiv \frac{1}{T_0}\int_1^2 DQ_{\text{I}}^0 = \int_1^2 \frac{DQ_{\text{I}}}{T_{\text{I}}} \quad \text{and} \quad \frac{Q_{\text{II}}^0}{T_0} \equiv \frac{1}{T_0}\int_1^2 DQ_{\text{II}}^0 = \int_1^2 \frac{DQ_{\text{II}}}{T_{\text{II}}}.$$

From the Kelvin-Planck principle it follows that W is necessarily less or equal zero, because there is only the single heat source R from which W is extracted. Invoking the first law of thermodynamics we get,

$$W = Q_{\text{I}}^0 - Q_{\text{II}}^0 \le 0 \Rightarrow \frac{Q_{\text{I}}^0}{T_0} \le \frac{Q_{\text{II}}^0}{T_0},$$

So that:

$$\int_1^2 \frac{DQ_{\text{I}}}{T_{\text{I}}} \le \int_1^2 \frac{DQ_{\text{II}}}{T_{\text{II}}}$$

Because the processes inside C and E are assumed reversible, all arrows can be reverted and in the very same way it is shown that:

$$\int_1^2 \frac{DQ_{\mathrm{II}}}{T_{\mathrm{II}}} \le \int_1^2 \frac{DQ_{\mathrm{I}}}{T_{\mathrm{I}}},$$

So that equation $\int_{1(I)}^{2} \frac{DQ}{T} = \int_{1(II)}^{2} \frac{DQ}{T}$. holds (with a slight change of notation, subscripts are transferred to the respective integral signs):

$$\int_{1(I)}^{2} \frac{DQ}{T} = \int_{1(II)}^{2} \frac{DQ}{T}.$$

Relation to Gibbs Free Energy and Enthalpy

The definition of Gibbs free energy is based on entropy as follows:

$$G = H - TS$$

Where all the thermodynamic properties except T are extensive and where:

G = Gibbs free energy

H = enthalpy

T = absolute temperature

S = entropy

A corresponding equation with all intensive properties (i.e., per unit of mass) can be written as follows:

$$g = h - Ts$$

Where,

g = specific Gibbs free energy

h = specific enthalpy

T = absolute temperature

S = specific entropy

Entropy of an Ideal Gas

The equation of state of one mole of an ideal gas is:

$$pV = RT$$

Where R is the molar gas constant, p the pressure, and V the volume of the gas. Note that the limit $T \to 0$ implies $V \to 0$—ideal-gas particles are of zero size.

The entropy of one mole of an ideal gas is a function of T and V and depends parametrically on the molar gas constant R and the molar heat capacity at constant volume, C_V,

$$S(T,V) = C_V \log(T) + R\log(V) + S_0 = R\log(T^{\frac{C_V}{R}} V) + S_0$$

Where S_0 is a constant independent of T, V, and p. From statistical thermodynamics it is known that for an atomic ideal gas $C_V = 3R/2$, so that the exponent of T becomes 3/2. For a diatomic ideal gas $C_V = 5R/2$ and for an ideal gas of arbitrarily shaped molecules $C_V = 3R$. In any case, for an ideal gas C_V is constant, independent of T, V, or p.

The expression for the ideal gas entropy is derived easily by substituting the ideal gas law (E1) into the following general differential equation for the entropy as function of T and V—valid for any thermodynamic system,

$$dS = \frac{C_V}{T} dT + \left(\frac{\partial p}{\partial T}\right)_V dV.$$

Integration gives:

$$\begin{aligned} \int_1^2 dS &= C_V \int_1^2 \frac{dT}{T} + R\int_1^2 \frac{dV}{V} \Rightarrow \\ S_2 - S_1 &= C_V \log(T_2) + R\log(V_2) - C_V \log(T_1) - R\log(V_1). \end{aligned}$$

Write,

$$S_0 \equiv S_1 - C_V \log(T_1) - R\log(V_2) \quad \text{and} \quad S_2 \equiv S, T_2 \equiv T, V_2 \equiv V$$

and the result follows.

Proof of Differential Equation for S(T,V)

The proof of the differential equation $dS = \frac{C_V}{T} dT + \left(\frac{\partial p}{\partial T}\right)_V dV$ follows by some typical classical thermodynamics calculus.

First, the internal energy at constant volume follows thus,

$$dU = \left(\frac{\partial U}{\partial T}\right)_V dT + \left(\frac{\partial U}{\partial V}\right)_T dV \underset{\text{constant } V}{\Rightarrow} dU = \left(\frac{\partial U}{\partial T}\right)_V dT.$$

The definition of heat capacity and the first law (DQ = dU + pdV, for constant volume: DQ = dU) give,

$$DQ \equiv C_V\, dT = dU = \left(\frac{\partial U}{\partial T}\right)_V dT$$

So that the heat capacity at constant volume is given by:

$$C_V = \left(\frac{\partial U}{\partial T}\right)_V .$$

The first and second law combined (TdS = dU + pdV) gives:

$$dS = \underbrace{\frac{C_V}{T}}_{\frac{\partial S}{\partial T}} dT + \underbrace{\frac{1}{T}\left[\left(\frac{\partial U}{\partial V}\right)_T + p\right]}_{\frac{\partial S}{\partial V}} dV.$$

From,

$$\frac{\partial}{\partial V}\frac{\partial S}{\partial T} = \frac{\partial}{\partial T}\frac{\partial S}{\partial V}$$

and

$$\frac{\partial}{\partial V}\frac{\partial S}{\partial T} = \frac{\partial}{\partial V}\frac{C_V}{T} = \frac{1}{T}\frac{\partial C_V}{\partial V} = \frac{1}{T}\frac{\partial^2 U}{\partial V \partial T}$$

and

$$\frac{\partial}{\partial T}\frac{\partial S}{\partial V} = \frac{\partial}{\partial T}\frac{1}{T}\left[\left(\frac{\partial U}{\partial V}\right)_T + p\right] = -\frac{1}{T^2}\left[\left(\frac{\partial U}{\partial V}\right)_T + p\right] + \frac{1}{T}\left[\left(\frac{\partial^2 U}{\partial T \partial V}\right) + \left(\frac{\partial p}{\partial T}\right)_V\right]$$

Follows:

$$0 = -\frac{1}{T^2}\left[\left(\frac{\partial U}{\partial V}\right)_T + p\right] + \frac{1}{T}\left(\frac{\partial p}{\partial T}\right)_V \Rightarrow \left(\frac{\partial U}{\partial V}\right)_T = -p + T\left(\frac{\partial p}{\partial T}\right)_V .$$

Substitute the very last equation into equation $dS = \underbrace{\frac{C_V}{T}}_{\frac{\partial S}{\partial T}} dT + \underbrace{\frac{1}{T}\left[\left(\frac{\partial U}{\partial V}\right)_T + p\right]}_{\frac{\partial S}{\partial V}} dV$, and the equation to be proved follows,

$$dS = \frac{C_V}{T} dT + \left(\frac{\partial p}{\partial T}\right)_V dV.$$

Entropy in Statistical Thermodynamics

In classical (phenomenological) thermodynamics it is not necessary to assume that matter consists of small particles (atoms or molecules). While this has the advantage of keeping classical thermodynamics transparent, not obscured by microscopic details, and universally valid, independent of the kind of molecules constituting the system, it has the disadvantage that it cannot predict the value of any parameters. For instance, the heat capacity of a monatomic ideal gas at constant volume C_V is equal to $3R/2$, where R is the molar gas constant. One needs a microscopic theory to find this simple result.

Before the 1920s the microscopic (molecular) theory of thermodynamics was based on classical (Newtonian) mechanics and on the kind of statistical arguments that were first introduced into physics by Maxwell and developed by Gibbs and Boltzmann. The branch of physics that tries to predict thermodynamic properties departing from molecular properties is known as statistical thermodynamics or statistical mechanics. Since the 1920s statistical thermodynamics is based usually on quantum mechanics.

The statistical mechanics expression for the entropy is,

$$S = -k_B \mathrm{Tr}[\hat{\rho} \log \hat{\rho}]$$

where the density operator $\hat{\rho}$ is given by:

$$\hat{\rho} = \frac{e^{-\hat{H}/(k_B T)}}{\mathrm{Tr}[e^{-\hat{H}/(k_B T)}]}.$$

Further k_B is Boltzmann's constant, $\hat{H}$ is the quantum mechanical energy operator of the total system (the energies of all particles plus their interactions), and the trace (Tr) of an operator is the sum of its diagonal matrix elements.

It will also be shown under which circumstance the entropy may be given by Boltzmann's celebrated equation:

$$S = k \log W.$$

Density Operator

John von Neumann introduced into quantum mechanics the density operator $\hat{\rho}$ (called "statistical operator" by von Neumann) for a system of which the state is only partially known. He considered the situation that certain real numbers p_m are known that correspond to a complete set of *orthonormal* quantum mechanical states (m = 0, 1, 2, ..., ∞). The quantity p_m is the probability that state $|m\rangle$ is occupied, or in other words, it is the percentage of systems in a (very large) ensemble of identical systems that are in the state $|m\rangle$. As is usual for probabilities, they are normalized to unity,

$$\sum_{m=0}^{\infty} p_m = 1.$$

The averaged value of a property with quantum mechanical operator $\hat{P}$ of a system described by the probabilities p_m is given by the ensemble average,

$$\langle\langle \hat{P} \rangle\rangle \equiv \sum_{m=0}^{\infty} p_m \langle m | \hat{P} | m \rangle,$$

where $\langle m | \hat{P} | m \rangle$ is the usual quantum mechanical expectation value.

The expression for $\langle\langle P \rangle\rangle$ can be written as a trace of an operator product. First define the density operator;

$$\hat{\rho} \equiv \sum_{n=0}^{\infty} |n\rangle p_n \langle n|,$$

Then it follows that:

$$\langle\langle \hat{P} \rangle\rangle = \mathrm{Tr}\left[\hat{P}\hat{\rho}\right].$$

Indeed,

$$\begin{aligned}\mathrm{Tr}\left[\hat{P}\hat{\rho}\right] &\equiv \sum_{m} \langle m|\hat{P}\hat{\rho}|m\rangle = \sum_{nm} \langle m|n\rangle p_n \langle n|\hat{P}|m\rangle = \sum_{nm} p_n \delta_{mn} \langle n|\hat{P}|m\rangle \\ &= \sum_{m} p_m \langle m|\hat{P}|m\rangle = \langle\langle \hat{P} \rangle\rangle,\end{aligned}$$

Where $\langle m \mid n \rangle = \delta_{mn}$, the Kronecker delta.

A density operator has unit trace:

$$\mathrm{Tr}\hat{\rho} = \sum_{mn} \langle m|n\rangle p_n \langle n|m\rangle = \sum_{n} p_n \sum_{m} \delta_{nm}\delta_{mn} = \sum_{n} p_n \delta_{nn} = \sum_{n} p_n = 1.$$

Closed Isothermal System

For a thermodynamic system of constant temperature (T), volume (V), and number of particles (N),one considers eigenstates of the energy operator $\widehat{H}$, the Hamiltonian of the *total* system,

$$\hat{H}|m\rangle = E_m|m\rangle$$

Assume that p_m is proportional to the Boltzmann factor, with the proportionality constant K determined by normalization,

$$p_m = Ke^{-E_m/(k_B T)} \quad \text{with} \quad K\sum_{m} e^{-E_m/(k_B T)} = 1 \Rightarrow K = \left[\sum_{m} e^{-E_m/(k_B T)}\right]^{-1}$$

Where k_B is the Boltzmann constant. It is common to designate the *partition function* of the system of constant T, N, and V by Q,

$$Q \equiv \sum_{m=0}^{\infty} e^{-E_m/(k_B T)}.$$

Hence, using that:

$$\langle m|e^{-\hat{H}/(k_B T)}|m\rangle = e^{-E_m/(k_B T)}$$

It is found,

$$\hat{\rho} = \frac{1}{Q}\sum_{m} |m\rangle\langle m| e^{-\hat{H}/(k_B T)} |m\rangle\langle m| = \frac{1}{Q}\sum_{mn} |m\rangle\langle m| e^{-\hat{H}/(k_B T)} |n\rangle\langle n| = \frac{\exp[-\hat{H}/(k_B T)]}{Q},$$

where it used that the set of states is complete—give rise to the following resolution of the identity operator,

$$\hat{1} = \sum_{m} |m\rangle\langle m| = \sum_{n} |n\rangle\langle n|.$$

In summary, the canonical ensemble average of a property with quantum mechanical operator $\widehat{P}$ is given by

$$\langle\langle \hat{P} \rangle\rangle = \mathrm{Tr}\left[\hat{P}\,\hat{}\,\right] = —\mathrm{Tr}\left[\hat{P}e^{\ \hat{H}/(k\ T)}\right].$$

Internal Energy

The quantum statistical expression for internal energy is:

$$U \equiv \langle\langle \hat{H} \rangle\rangle = \mathrm{Tr}\left[\hat{H}\hat{\rho}\right] = \frac{1}{Q}\mathrm{Tr}\left[\hat{H}e^{-\hat{H}/(k_B T)}\right].$$

From,

$$\log\hat{\rho} = \log\left[e^{-\hat{H}/(k_B T)}/Q\right] = -\hat{H}/(k_B T) - \hat{1}\log Q$$

Follows,

$$\hat{H} = -k_B T\left(\log\hat{\rho} + \hat{1}\log Q\right).$$

The quantum statistical expression for the internal energy U becomes:

$$U = \mathrm{Tr}\left[-k_B T\left(\log\hat{\rho} + \hat{1}\log Q\right)\hat{\rho}\right] = -T\,k_B\,\mathrm{Tr}[\hat{\rho}\log\hat{\rho}] - T\,k_B\log(Q),$$

Where it is used that a scalar may be taken of the trace and that the density operator is of unit trace.

In classical thermodynamics the internal energy is related to the entropy S and the Helmholtz free energy A by,

$$U = T\,S + A.$$

Define,

$$\hat{S} \equiv -k_B\log\hat{\rho}, \quad \hat{A} \equiv -k_B\,T\log(Q)\,\hat{1}$$

And accordingly,

$$S \equiv \langle\langle \hat{S} \rangle\rangle = \mathrm{Tr}[\hat{S}\hat{\rho}] = -k_\mathrm{B}\,\mathrm{Tr}[\hat{\rho}\log\hat{\rho}]$$

and

$$A \equiv \langle\langle \hat{A} \rangle\rangle = -\mathrm{Tr}[\hat{A}\hat{\rho}] = -k_\mathrm{B}\,T\log(Q)\mathrm{Tr}[\hat{\rho}] = -k_\mathrm{B}\,T\log(Q).$$

As a result,

$$U = TS + A = -k_\mathrm{B}\,T\,\mathrm{Tr}[\hat{\rho}\log\hat{\rho}] - k_\mathrm{B}\,T\log(Q),$$

Which agrees with the quantum statistical expression for U, which in turn means that the definitions $\hat{S} \equiv -k_\mathrm{B}\log\hat{\rho}$, $\hat{A} \equiv -k_\mathrm{B}\,T\log(Q)\hat{1}$ of the entropy operator and Helmholtz free energy operator are consistent.

Note that neither the entropy nor the free energy are given by an ordinary quantum mechanical operator, both depend on the temperature through the partition function Q. Furthermore Q is defined as a trace:

$$Q = \mathrm{Tr}[e^{-\hat{H}/(k_\mathrm{B}T)}]$$

and thus samples the whole (Hilbert) space containing the state vectors $|\,m\,\rangle$. Almost all quantum mechanical operators that represent observable (physical) quantities have a classical (electromagnetic or mechanical) counterpart. Clearly the entropy operator lacks such a parallel definition, and this is probably the main reason why entropy is a concept that is difficult to comprehend.

Boltzmann's Formula for Entropy

Let us consider an isolated system (constant *U*, *V*, and *N*). Traces are taken only over states with energy *U*. Let there be $\Omega(U, V, N)$ of these states. This is in general a very large number, for instance for one mole of a monatomic ideal gas consisting of $N = N_\mathrm{A} \approx 10^{23}$ (Avogadro's number) it holds that,

$$\Omega(U,V,N) = \left[\left(\frac{2\pi m k_\mathrm{B}T}{h^2}\right)^{3/2}\frac{Ve^{5/2}}{N^{5/2}}\right]^N \approx e^N \approx 10^{10^{23}}.$$

Here m is the mass of an atom, h is Planck's constant, V is the volume of the vessel containing the gas, and $e \approx 2.7$.

The sum in the partition function shrinks to a sum over Ω states of energy U, hence:

$$Q = \mathrm{Tr}\left[e^{-\hat{H}/(k_\mathrm{B}T)}\right] = \Omega(U,V,N)e^{-U/(k_\mathrm{B}T)}$$

Likewise,

$$S = -k_\mathrm{B}\mathrm{Tr}\rho\log\rho = -k_\mathrm{B}\Omega\frac{e^{-U/(k_\mathrm{B}T)}}{Q}\log\left(\frac{e^{-U/(k_\mathrm{B}T)}}{Q}\right) = -k_\mathrm{B}\log\frac{1}{\Omega},$$

So that Boltzmann's celebrated equation follows:

$$S = k_B \log \Omega(U,V,N)$$

From the previous expression for Ω follows an expression for the entropy of a monatomic ideal gas as a function of T and V,

$$S = Nk_B \log(V T^{3/2}) + S_0 \quad \text{with} \quad S_0 = Nk_B \log\left[\left(\frac{2\pi m k_B}{h^2}\right)^{3/2} \frac{e^{5/2}}{N^{5/2}}\right].$$

Recalling that $N_A k_B \equiv R$ and $C_V = 3/2\ R$ one sees that this is the formula encountered above [between equation, $pV = RT$, and $dS = \frac{C_V}{T} dT + \left(\frac{\partial p}{\partial T}\right)_V dV$], but this time with an explicit expression for S_o.

Boltzmann's equation is derived as an average over an ensemble consisting of identical systems of constant energy, number of particles, and volume; such an ensemble is known as a *microcanonical ensemble*. However, it can be shown that energy fluctuations around the mean energy in a canonical ensemble (constant *T*) are extremely small, so that taking the trace over only the states of mean energy is a very good approximation. In other words, although Boltzmann's formula does not hold formally for a canonical ensemble, in practice it is a *very* good approximation, also for isothermal systems.

Entropy as Disorder

In common parlance the term *entropy* is used for lack of order and gradual decline into disorder. One can find in many introductory physics texts the statement that entropy is a measure for the degree of randomness in a system.

The origin of these statements is Boltzmann's 1877 equation $S=k_B \log\Omega$. The third law of thermodynamics states the following: when $T \rightarrow 0$ the number of accessible states Ω goes to unity, and the entropy *S* goes to zero. That is, if one interprets entropy as randomness, then at zero K there is no disorder whatsoever, matter is in complete order. Clearly, this low-temperature limit supports the intuitive notion of entropy as a measure of chaos.

Ω gives the number of quantum states accessible to a system. It can be argued that the more quantum states are available to a system, the greater the complexity of the system. If one equates complexity with randomness, as is often done in this context, it confirms the notion of entropy as a measure of disorder. The second law of thermodynamics, which states that a spontaneous process in an isolated system strives toward maximum entropy, can be interpreted as the tendency of the universe to become more and more chaotic.

However, the view of entropy as disorder, as a measure of chaos, is disputed. For instance, Lambert contends that entropy is a "measure for energy dispersal". If one reads "energy dispersal" as heat divided by temperature, this is true by the classical (phenomenological) definition of entropy. Lambert states that from a molecular point of view, entropy increases when more microstates become available to the system (i.e., Ω increases) and the energy is dispersed over the greater number of accessible microstates. Lambert argues further that the view of entropy as disorder, is "so misleading as actually to be a failure-prone crutch".

If one rejects completely the idea of entropy as randomness, one discards a convenient mnemonic device. Generations of physicists and chemists have remembered that a gas contains more entropy than a crystal, "because a gas is more chaotic than a crystal". This is easier to remember than "because the gas has more microstates to its disposal and its energy is dispersed over these larger number of microstates", although the latter statement is the more correct one.

Entropy as Function of Aggregation State

As just stated, the entropy of a mole of pure substance changes as follows:

$$S_{gas} > S_{liq} > S_{sol}$$

Which agrees with our intuition that a gas is more chaotic than a liquid, which again is more chaotic than a solid.

As an illustration of this point, consider one mole of water (H_2O) at a pressure of 1 bar (≈ 1 atmosphere). Experimentally, the enthalpy of fusion ΔH_f is 6.01 kJ/mol and the enthalpy of vaporization ΔH_v is 40.72 kJ/mol. Remember that enthalpy is heat added/extracted reversibly at constant pressure (in this case 1 bar) to achieve the change of aggregation state. Further the change of aggregation state occurs at constant temperature, so that,

$$\Delta S_f = \frac{\Delta H_f}{T_f} \quad \text{and} \quad \Delta S_v = \frac{\Delta H_v}{T_v}.$$

For water T_f = 0 °C = 273.15 K and T_v = 100 °C = 373.15 K. Hence

$$\Delta S_f = 22.0\,\text{J/(mol K)} \quad \text{and} \quad \Delta S_v = 109.1\,\text{J/(mol K)}.$$

Summarizing, in units J/(mol K) a mole of liquid water contains 22.0 more entropy than a mole of ice (both at 0 °C); a mole of gas (steam at 100 °C) contains 109.1 more entropy than a mole of liquid water at boiling temperature.

The Increase of Entropy Principle

Entropy change of a closed system during an irreversible process is greater that the integral of δQ / T evaluated for the process. In the limiting case of a reversible process, they become equal.

$$dS \geq \frac{\delta Q}{T}$$

The entropy generated during a process is called entropy generation, and is denoted by S_{gen},

$$\Delta S = S_2 - S_1 = \int_1^2 \frac{\delta Q}{T} + S_{gen}$$

Note that the entropy generation S_{gen} is always a positive quantity or zero (reversible process). Its value depends on the process, thus it is not a property of a system.

The entropy of an isolated system during a process always increases, or in the limiting case of a reversible process remains constant (it never decreases). This is known as the increase of entropy principle.

The entropy change of a system or its surroundings can be negative; but entropy generation cannot.

$$S_{gen} = \begin{cases} > 0 & \text{irreversible process} \\ = 0 & \text{reversible process} \\ < 0 & \text{imposible process} \end{cases}$$

1- A process must proceeds in the direction that complies with the increase of entropy principle, $S_{gen} > 0$. A process that violates this principle is impossible.

2- Entropy is a non-conserved property, and there is no such thing as the conservation of entropy. Therefore, the entropy of universe is continuously increasing.

3- The performance of engineering systems is degraded by the presence of irreversibility. The entropy generation is a measure of the magnitudes of the irreversibilities present during the process.

Entropy Balance

Entropy is a measure of molecular disorder or randomness of a system, and the second law states that entropy can be created but it cannot be destroyed.

The increase of entropy principle is expressed as:

Entropy change = Entropy transfer + Entropy generation

$$\Delta S_{system} = S_{transfer} + S_{gen}$$

This is called the entropy balance.

Entropy Change

The entropy balance is easier to apply that energy balance, since unlike energy (which has many forms such as heat and work) entropy has only one form. The entropy change for a system during a process is:

Entropy change = Entropy at final state - Entropy at initial state

$$\Delta S_{system} = S_{final} - S_{initial}$$

Therefore, the entropy change of a system is zero if the state of the system does not change during the process. For example entropy change of steady flow devices such as nozzles, compressors, turbines, pumps, and heat exchangers is zero during steady operation.

Mechanisms of Entropy Transfer

Entropy can be transferred to or from a system in two forms: heat transfer and mass flow. Thus, the entropy transfer for an adiabatic closed system is zero.

Heat Transfer: heat is a form of disorganized energy and some disorganization (entropy) will flow with heat. Heat rejection is the only way that the entropy of a fixed mass can be decreased. The ratio of the heat transfer Q/ T (absolute temperature) at a location is called entropy flow or entropy transfer.

$$\text{Entropy transfer with heat (T = const.)}\ S_{heat} = \frac{Q}{T}$$

Since T (in Kelvin) is always positive, the direction of entropy transfer is the same of the direction of heat transfer.

When two systems are in contact, the entropy transfer from warmer system is equal to the entropy transfer to the colder system since the boundary has no thickness and occupies no volume.

Note that work is entropy-free, and no entropy is transferred with work.

Mass Flow: mass contains entropy as well as energy, both entropy and energy contents of a system are proportional to the mass. When a mass in the amount of m enters or leaves a system, entropy in the amount of ms (s is the specific entropy) accompanies it.

Entropy Balance for a Closed System

A closed system includes no mass flow across its boundaries, and the entropy change is simply the difference between the initial and final entropies of the system.

The entropy change of a closed system is due to the entropy transfer accompanying heat transfer and the entropy generation within the system boundaries:

Entropy change of the system = Entropy transfer with heat + Entropy generation

$$S_2 - S_1 = \sum \frac{Q_k}{T_k} + S_{gen}$$

Therefore, for an adiabatic closed system, we have:

$$\Delta S_{adiabatic} = S_{gen}$$

For an internally reversible adiabatic process $\Delta S = 0$, because $S_{gen} = 0$.

The total entropy generated during a process can be determined by applying the entropy balance to an extended system that includes both the system and its immediate surroundings where external irreversibility might be occurring.

Example of Entropy Balance for a Closed System

Saturated liquid water at 100 C is contained in a piston-cylinder assembly. The water undergoes a process to the corresponding saturated vapor state, during which the piston moves freely in the cylinder. There is no heat transfer with the surroundings. If the change of state is brought about by the action of a paddle wheel, determine the network per unit mass, in kJ/kg, and the amount of entropy produced per unit mass, in kJ/kg.K.

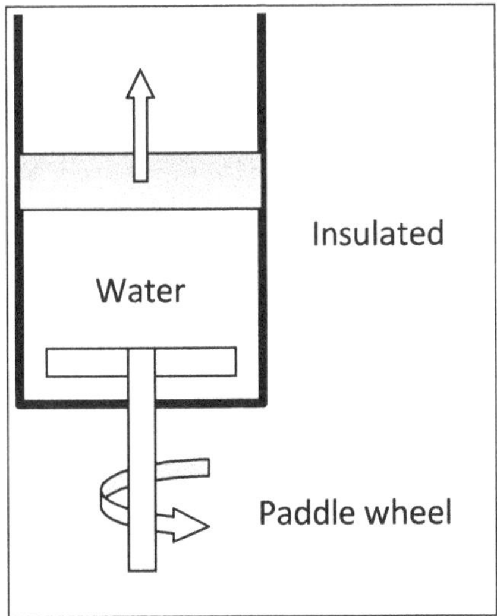

Assumptions:

- The water in the piston-cylinder assembly is a closed system.
- There is no heat transfer with the surroundings.
- The system is at an equilibrium state initially and finally. $\Delta PE = \Delta KE = 0$.

Solution:

The network can be calculated by using the law:

$$\Delta U + \Delta KE + \Delta PE = Q - W$$

That is simplifies to: $\Delta U = - W$

On a unit mass basis, the energy balance becomes:

$$W / m = -(u_g - u_f)$$

$$W / m = -2087.6 \text{ kJ/kg}$$

The negative sign indicates that the work input by the stirring is greater than the work done by the water as it expands.

Using an entropy balance, the amount of entropy produced can be found. Since there is no heat transfer,

$$\Delta S = \underbrace{\int_1^2 \left(\frac{\delta Q}{T}\right)}_{0} + S_{gen} = S_{gen}$$

On a unit mass basis, this becomes:

$$S_{gen} / m = s_g - s_f$$

$$S_{gen} / m = 6.048\ kJ / kg.K$$

Entropy Balance for a Control Volume

In addition to methods discussed for closed system, the entropy can be exchanged through mass flows across the boundaries of the control volume.

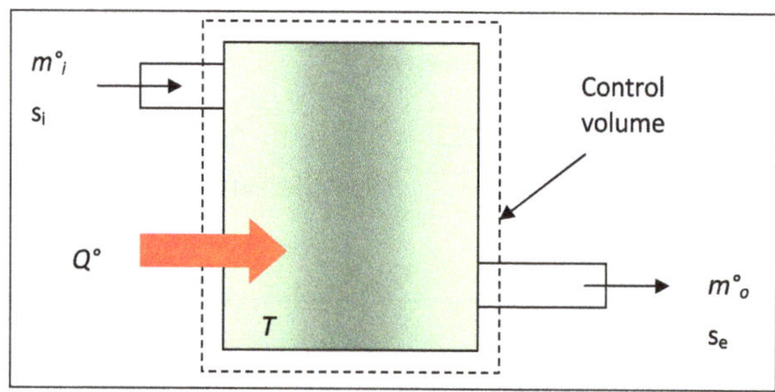

The entropy balance in the rate form for a control volume becomes:

$$\frac{dS_{CV}}{dt} = \sum \frac{Q_k^{\bullet}}{T_k} + \sum m_i^{\bullet} s_i - \sum m_e^{\bullet} s_e + S_{gen,CV}^{\bullet}$$

For a steady-state steady-flow process, it simplifies to:

$$S_{gen,CV}^{\bullet} = \sum m_e^{\bullet} s_e - \sum m_i^{\bullet} s_i - \sum \frac{Q_k^{\bullet}}{T_k}$$

Example of Entropy Balance for a CV

Steam enters a turbine with a pressure of 3 MPa, a temperature of 400 °C, and a velocity of 160 m/s. Saturated vapor at 100 °C exits with a velocity of 100 m/s. At steady-state, the turbine develops work equal to 540 kJ/kg. Heat transfer between the turbine and its surroundings occur at an average outer surface temperature of 350 K. Determine the rate at which entropy is produced within the turbine per kg of steam flowing, in kJ/kg.K. Neglect the change in potential energy between inlet and exit.

Assumptions:

- Steady state operation in CV. ΔPE = 0.
- Turbine outer surface is at a specified average temperature.

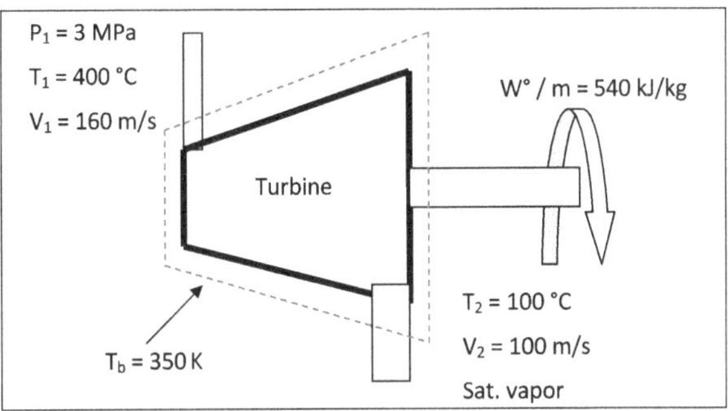

From the mass balance, we know that $m^{\circ} = m^{\circ}_1 = m^{\circ}_2$

Since the process is steady-state, one can write:

$$0 = \sum \frac{Q_k^{\bullet}}{T_k} + m^{\bullet}\left(s_i - s_e\right) + S^{\bullet}_{gen,CV}$$

The heat transfer occurs at Tb = 350 K, the first term of the right hand side of the entropy balance reduces to Q°/T_b:

$$\frac{S^{\bullet}_{gen,\,CV}}{m^{\bullet}} = -\frac{Q_k^{\bullet}}{m^{\bullet}T_k} + \left(S_2 - S_1\right)$$

We need to calculate the rate of heat transfer. The first law (energy balance) can be used to find the heat transfer rate. Combining the mass balance and the first law, one finds:

$$\frac{Q^{\bullet}_{CV}}{m^{\bullet}} = \frac{W^{\bullet}_{CV}}{m^{\bullet}} + \left(h_2 - h_1\right) + \left(\frac{V_2^2 - V_1^2}{2}\right)$$

h_1 = 3230.9 kJ/kg, and From A-4 h_2 = 2676.1 kJ/kg. After substitution, and converting the units, one finds:

$$\frac{Q^{\bullet}_{CV}}{m^{\bullet}} = -22.6\ kJ/kg$$

s_2 = 7.3549 kJ/kg. K and s_1 = 6.9212 kJ/kg. K. Inserting values into the expression for entropy production:

$$\frac{S^{\bullet}_{gen,CV}}{m^{\bullet}} = -\frac{Q_k^{\bullet}}{m^{\bullet}\,T_k} + \left(S_2 - S_1\right) = 0.4983\ kJ/kg.K$$

Evaluation of Entropy Change

The differential form of the conservation of energy for a closed system (fixed mass) for an internally reversible process is:

$$\delta Q_{\text{int, rev}} - \delta W_{\text{int, rev}} = dU$$

where,

$$\delta Q_{int,rev} = TdS$$

$$\delta W_{int,rev} = PdV$$

Thus,

$$TdS = dU + PdV$$

Or, per unit mass

$$Tds = du + Pdv$$

This is called the first Gibbs equation.

From the definition of enthalpy, h = u + Pv, one can find:

$$h = u + Pv \rightarrow dh = du + Pdv + vdP$$

Eliminating du from the first Gibbs equation, one finds the second Gibbs equation:

$$Tds = dh - vdP$$

Explicit relations for differential changes in entropy can be obtained from Gibbs equations:

$$dS = \frac{du}{T} + \frac{Pdv}{T}$$

$$dS = \frac{dh}{T} - \frac{vdP}{T}$$

To calculate the entropy change, we must know the relationship between du or dh and temperature.

Entropy Change of Solids and Liquids

Solids and liquids can be assumed as incompressible substances since their volumes remains essentially constant during a process. Thus, the first Gibbs equation becomes:

$$ds = \frac{du}{T} = \frac{cdT}{T}$$

$$s_2 - s_1 = \int_1^2 c(T)\frac{dT}{T}$$

Assuming an averaged value for specific heat, one obtains:

$$s_2 - s_1 = c_{ave} \text{ In } \frac{T_2}{T_1}$$

Note that the entropy change of an incompressible substance is only a function of temperature. Therefore, for an isentropic process where $s_2 = s_1$, one can find:

$$T_2 = T_1$$

Entropy Change of Ideal Gas

The entropy change of an ideal gas can be obtained, by substituting du = c_v dT and P = RT/v into Gibbs equation.

$$ds = c_v \frac{dT}{T} + R\frac{dv}{v}$$

$$s_2 - s_1 = \int_1^2 c_v(T)\frac{dT}{T} + R\ In\frac{v_2}{v_1}$$

Assuming averaged values for specific heats, one obtains:

$$s_2 - s_1 = c_{v,ave}\ In\ \frac{T_2}{T_1} + R\,In\frac{v_2}{v_1}\ \frac{kJ}{kg.K}$$

$$s_2 - s_1 = c_{p,ave}\ In\ \frac{T_2}{T_1} + R\,In\frac{P_2}{P_1}\ \frac{kJ}{kg.K}$$

For isentropic processes of ideal gases, the following relationships can be found by setting ds = 0,

$$\text{In}\frac{T_2}{T_1} = -\frac{R}{C_v}\text{In}\frac{v_2}{v_1}$$

$$\text{In}\frac{T_2}{T_1} = \text{In}\left(\frac{v_1}{v_2}\right)^{\frac{R}{C_v}} \quad or \quad \left(\frac{T_2}{T_1}\right) = \left(\frac{v_1}{v_2}\right)^{k-1} \quad \text{isentropic process}$$

Since $R = c_p - c_{v,}$ $k = c_p / c_v$ and thus $R / c_v = k - 1$.

In a similar manner, one finds:

$$\left(\frac{T_2}{T_1}\right) = \left(\frac{P_2}{P_1}\right)^{(k-1)/k} \quad \text{isentropic process}$$

$$\left(\frac{P_2}{P_1}\right) = \left(\frac{v_1}{v_2}\right)^{k} \quad \text{isentropic process}$$

These equations can be expressed in the following compact forms:

$$Tv^{k-1} = \text{constant}$$

$$TP^{(1-k)/k} = \text{constant}$$

$$Pv^{k} = \text{constant}$$

The specific ratio k, varies with temperature, and in isentropic relations above an average k value should be used.

Example of Isentropic Process of Ideal Gas

A rigid, well-insulated tank is filled initially with 5 kg of air at pressure 500 kPa and a temperature 500 K. A leak develops, and air slowly escapes until the pressure of the air remaining in the tank is 100 kPa. Using the ideal gas model, determine the amount of mass remaining in the tank and its temperature.

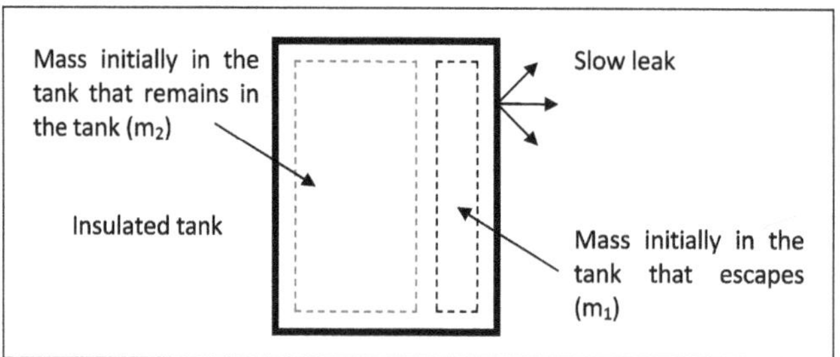

Assumptions:

- As shown in the figure, the closed system is the mass initially in the tank that remains in the tank.
- There is no significant heat transfer between the system and its surroundings.
- Irreversibilities within the tank can be ignored as the air slowly escapes.

Solutions:

Using the ideal gas equation of state, the mass initially in the tank that remains in the tank at the end of process is:

$$\left.\begin{aligned} m_2 &= \frac{P_2 V}{RT_2} \\ m_1 &= \frac{P_1 V}{RT_1} \end{aligned}\right\} \rightarrow m_2 = \left(\frac{P_2}{P_1}\right)\left(\frac{T_1}{T_2}\right) m_1$$

Since the volume of the tank V remains constant during the process. We need to find the final temperature T_2. For the closed system under consideration (m_1), there are no irreversibilities, and no heat transfer. Accordingly, it is an isentropic process, and thus the isentropic relationships can be used:

$$\frac{T_2}{T_1} = \left(\frac{P_2}{P_1}\right)^{(k-1)/k} \qquad T_2 = T_1\left(\frac{P_2}{P_1}\right)^{(k-1)/k}$$

With a constant k = 1.4 for air, after substituting values, one finds:

$T_2 = 315.55$ K

Finally, inserting values into the expression for system mass

$$m_2 = (100/500)\,(500/315.55)\,(5\text{ kg}) = 1.58\text{ kg}$$

Second Law of Thermodynamics in Terms of Entropy

The Second Law of Thermodynamics states that the state of entropy of the entire universe, as an isolated system, will always increase over time. The second law also states that the changes in the entropy in the universe can never be negative.

To understand why entropy increases and decreases, it is important to recognize that two changes in entropy have to considered at all times. The entropy change of the surroundings and the entropy change of the system itself. Given the entropy change of the universe is equivalent to the sums of the changes in entropy of the system and surroundings:

$$\Delta S_{univ} = \Delta S_{sys} + \Delta S_{surr} = \frac{q_{sys}}{T} + \frac{q_{surr}}{T}$$

In an isothermal reversible expansion, the heat q absorbed by the system from the surroundings is:

$$qrev = nRT \text{ In} \frac{V_2}{V_1}$$

Since the heat absorbed by the system is the amount lost by the surroundings, $q_{sys} = -q_{surr}$. Therefore, for a truly reversible process, the entropy change is:

$$\Delta S_{univ} = \frac{nRT \text{ In} \frac{V_2}{V_1}}{T} + \frac{-nRT \text{ In} \frac{V_2}{V_1}}{T} = 0$$

If the process is irreversible however, the entropy change is,

$$\Delta S_{univ} = \frac{nRT \text{ In} \frac{V_2}{V_1}}{T} > 0$$

If we put the two equations for ΔS_{univ} together for both types of processes, we are left with the second law of thermodynamics,

$$\Delta S_{univ} = \Delta S_{sys} + \Delta S_{surr} \geq 0$$

where ΔS_{univ} equals zero for a truly reversible process and is greater than zero for an irreversible process. In reality, however, truly reversible processes never happen (or will take an infinitely long time to happen), so it is safe to say all thermodynamic processes we encounter everyday are irreversible in the direction they occur.

The second law of thermodynamics can also be stated that "all spontaneous processes produce an increase in the entropy of the universe".

Clausius Theorem

According to Clausius' theorem a reversible path or process or line could be replaced by two reversible adiabatic processes or lines and one reversible isothermal process or line.

In the following figure, a system is at equilibrium state i.e. at point i and system reaches to another equilibrium state i.e. at point f by following the reversible path i-f.

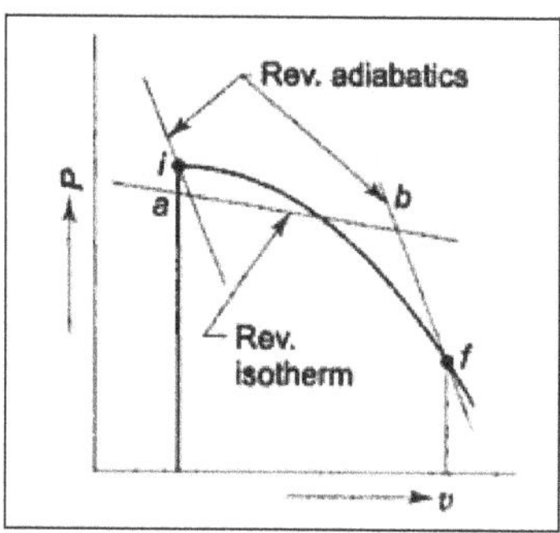

According to Clausius' theorem reversible process i.e. i-f could be replaced by two reversible adiabatic processes i.e. i-a and b-f and one reversible isothermal process i.e. a-b.

- Process i-f: Reversible process.
- Processes i-a and b-f: Reversible adiabatic processes.
- Process a-b: Reversible isothermal process.

The replacement will be done in such a way that the area below the reversible path i-f must be equal to the area below i-a-b-f. Mathematically:

Area below i-f = Area below i-a-b-f

Let us apply the "first law of thermodynamics" for following path or process:

Process: i-f,

$$Q_{if} = W_{if} + U_f - U_i$$

Process: i-a-b-f,

$$Q_{iabf} = W_{iabf} + U_f - U_i$$

As we have seen above that the area below the reversible path i-f must be equal to the area below i-a-b-f or $W_{if} = W_{iabf}$.

Therefore, we will have:

$$Q_{if} = Q_{iabf}$$

$$Q_{if} = Q_{ia} + Q_{ab} + Q_{bf}$$

As we know that i-a and b-f are reversible adiabatic processes and therefore $Q_{ia} = Q_{bf} = 0$.

$$Q_{if} = Q_{ab}$$

From above expression, we can say that heat transferred during the reversible process i-f will be equal to the heat transferred during the reversible isothermal process a-b.

Therefore we can say that any reversible process or path could be replaced by a reversible zigzag path with same end states and this reversible zigzag path will have two reversible adiabatic paths and one reversible isothermal path.

This replacement will be done in such a way that heat transferred during the original reversible process will be equal to the heat transferred during the reversible isothermal process.

We were discussing above single reversible process, now we will analyze here one complete reversible cycle as shown in figure. Let the reversible cycle is divided into large number of strips and these strips will indicate the reversible adiabatic lines as shown in figure. These strips are closed on bottom and tops with the help of reversible isothermal lines.

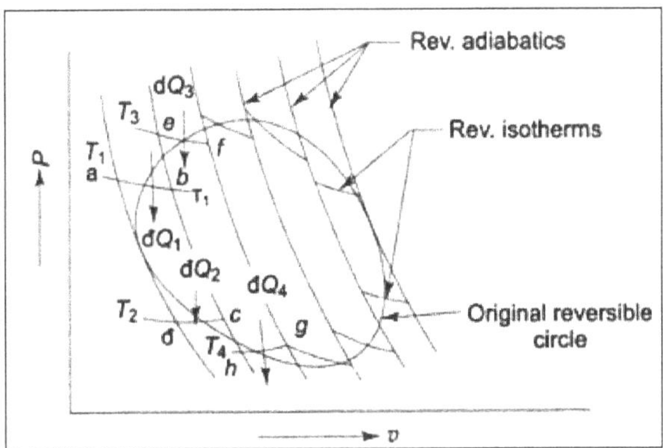

We can see here that original reversible cycle will be divided here into numbers of small-small Carnot's cycle as shown in figure.

Let us focus over here the Carnot's cycle abcd, dQ_1 heat is absorbed reversibly at temperature T_1 and dQ_2 heat is rejected reversibly at temperature T_2.

$$dQ_1 / T_1 = dQ_2 / T_2$$

Let us consider the sign convention and we will take heat absorption as positive and heat rejection as negative.

$$dQ_1 / T_1 + dQ_2 / T_2 = 0$$

In similar way for Carnot's cycle efgh, dQ_3 heat is absorbed reversibly at temperature T_3 and dQ_4 heat is rejected reversibly at temperature T_4.

$$dQ_3 / T_3 + dQ_4 / T_4 = 0$$

Similarly for complete reversible cycle, we will have following equation:

$$\frac{\delta Q_1}{T_1}+\frac{\delta Q_2}{T_2}+\frac{\delta Q_3}{T_3}+\frac{\delta Q_4}{T_4}+\ldots=0$$

$$\oint_R \frac{\delta Q}{T}=0$$

The cyclic integral of dQ/T for a reversible process will be zero and that is the mathematical expression of Clausius theorem. We must note it here that above equation will be only valid for a reversible cycle.

Clausius Inequality

The Clausius inequality provides a mathematical statement of the second law, which is a precursor to second law statements involving entropy. German physicist RJE Clausius, one of the founders of thermodynamics, stated,

$$\oint(\delta Q/T)\le 0$$

where the integral symbol $\oint$ shows the integration should be done for the entire system. The cyclic integral of $\delta Q/T$ is always less than or equal to zero. The system undergoes only reversible processes (or cycles) if the cyclic integral equals zero, and irreversible processes (or cycles) if it is less than zero.

Equation $\oint(\delta Q/T)\le 0$ can be expressed without the inequality as:

$$S_{gen}=-\oint(\delta Q/T)$$

Where,

$$S_{gen}=\Delta S_{total}=\Delta S_{sys}+\Delta S_{surr}$$

The quantity S_{gen} is the entropy generation associated with a process or cycle, due to irreversibilities. The following are cases for values of S_{gen}:

- S_{gen} = 0 for a reversible process
- S_{gen} > 0 for an irreversible process
- S_{gen} < 0 for no process (i.e., negative values for S_{gen} are not possible)

Consequently, one can write for a reversible process,

$$\Delta S_{sys}=(Q/T)_{rev} \text{ and } \Delta S_{surr}=-(Q/T)_{rev}$$

For an irreversible process,

$$\Delta S_{sys} > (Q/T)_{surr}$$

Due to entropy generation within the system as a result of internal irreversibilities. Hence, although the change in entropy of the system and its surroundings may individually increase, decrease or remain constant, the total entropy change or the total entropy generation cannot be less than zero for any process.

Temperature-Entropy Diagrams

In general, the phases of a substance and the relationships between its properties are most commonly shown on property diagrams. A large number of different properties have been defined, and there are some dependencies between properties.

A Temperature-entropy diagram (T-s diagram) is the type of diagram most frequently used to analyze energy transfer system cycles. It is used in thermodynamics to visualize changes to temperature and specific entropy during a thermodynamic process or cycle.

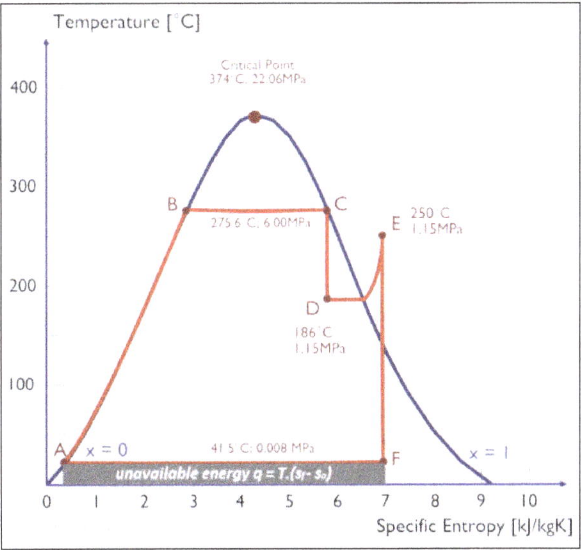

T-s diagram of Rankine Cycle.

This is because the work done by or on the system and the heat added to or removed from the system can be visualized on the T-s diagram. By the definition of entropy, the heat transferred to or from a system equals the area under the T-s curve of the process:

$$dQ = TdS$$

An isentropic process is depicted as a vertical line on a T-s diagram, whereas an isothermal process is a horizontal line. In an idealized state, compression is a pump, compression in a compressor and expansion in a turbine are isentropic processes. Therefore it is very useful in power engineering, because these devices are used in thermodynamic cycles of power plants.

Note that, the isentropic assumptions are only applicable with ideal cycles. Real thermodynamic cycles have inherent energy losses due to inefficiency of compressors and turbines.

Specific Entropy of Wet Steam

For pure substances like steam the specific entropy is included in the steam tables similar to specific volume, specific internal energy and specific enthalpy.

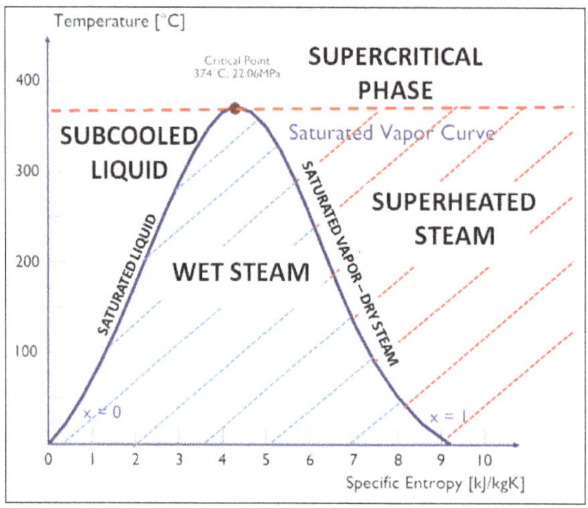

The specific entropy of saturated liquid water (x = 0) and dry steam (x = 1) can be picked from steam tables. In case of wet steam, the actual entropy can be calculated with the vapor quality, x, and the specific entropies of saturated liquid water and dry steam:

$$s_{wet} = s_s x + (1 - x) s_l$$

Where,

s_{wet} üüüüüüüüüüüü $(J \quad kg\ K)$

$s_s =$ entropy of "dry" steam $(J / kg\ K)$

$s_l =$ entropy of saturated liquid water $(J / kg\ K)$

Phase Transition

Phases are states of matter characterized by distinct macroscopic properties. Typical phases are liquid, solid and gas. Other important phases are superconducting and magnetic states.

First and second order phase transitions. States of matter come with their stability regions, the phase diagram. The properties of the microscopic state change by definition at the phase boundary. This change is:

$$\left.\begin{matrix}\text{discontinuous}\\ \text{continuous}\end{matrix}\right\} \text{ for a } \left\{\begin{matrix}\text{first order}\\ \text{second order}\end{matrix}\right\} \text{ phase transition}$$

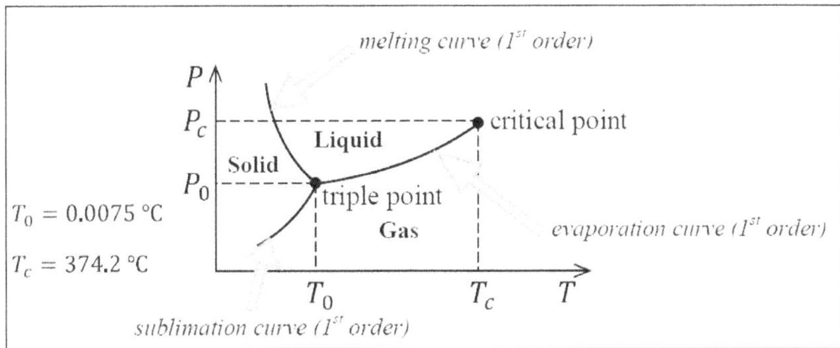

The appropriate variables for phase diagram of water are the pressure P and the temperature T.

- Critical Point : The first-order phase boundary between gas and liquid becomes second order right at the critical point. The two phases have then equal densities and specific entropies (entropy per particle). There is no critical point for the liquid-solid transition.
- Triple Point : The point at which gas, liquid and solid coexist.

Skating on ice: The melting curve of water has negative slope. Ice does hence melt at constant temperature $T < T_0$ when increasing the pressure P. This happens during skating on ice.

First-order Phase Transition

When water starts boiling, it undergoes a phase transition from a liquid to a gas phase. For both phases independently, the equation of state is a well-defined regular function, continuous, with continuous derivatives. However, while going from liquid to gas one function "abruptly" changes to the other function. Such a transition is of first-order.

Gibbs Enthalpy: Phase transitions in the P – T phase diagram are described by the Gibbs enthalpy G(T, P, N), which is itself a function of the pressure P and of the temperature T. G(T, P, N) changes continuously across the phase boundary when the transition is of first order. The entropy S and volume V, which are given by the derivatives:

$$S = -\left(\frac{\partial G}{\partial T}\right)_P, \quad V = \left(\frac{\partial G}{\partial P}\right)_T, \quad dG = -SdT + VdP + \mu dN$$

of the Gibbs potential, are in contrast discontinuous.

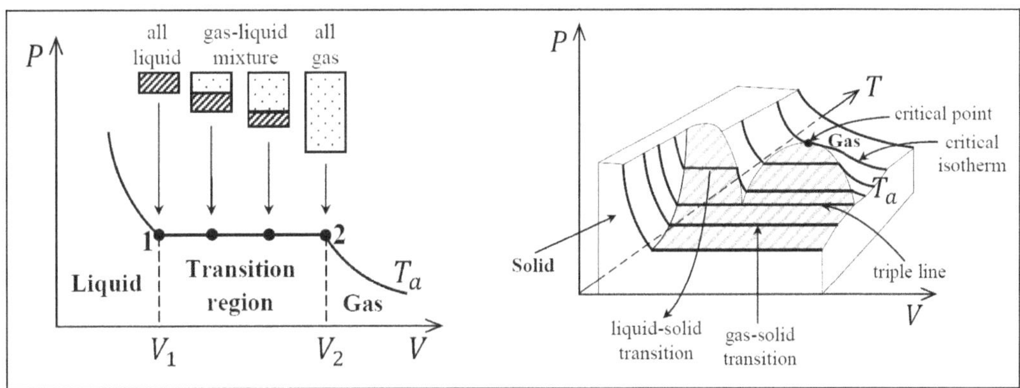

Latent Heat: Let us consider an instead of the P – T the P – V diagram, which is a projection of the equation of state for water.

Two phases 1 and 2 coexisting at a temperature T_0 have different entropies S_1 and S_2. The system must therefore absorb or release heat, the latent heat ΔQ_L,

$$\Delta Q_L = T_0 (S_2 - S_1),$$

During a phase transition of first order.

Condition for Phase Coexistence

We consider a system composed by one species of particles at given conditions of P and T (constant). Let us assume coexistence of two phases in our system. In this case, the,

Gibbs potential is the sum of Gibbs potentials of the two phases,

$$G(T,P,N) = G_1(T,P,N_1) + G_2(T,P,N_2)$$
$$= N_1\mu_1(T,P) + N_2\mu_2(T,P)$$

Where we have used the Gibbs-Duhem relation. Ni and μi are the number of particles and the chemical potential in phase i and $N = N_1 + N_2$.

Principle of minimal Gibbs energy: The principle of minimal Gibbs energy states that Gibbs potential has to be at its minimum, i.e. dG = 0, when P, T and $N = N_1 + N_2$ are fixed. G has to be therefore in equilibrium with respect to particle transfer from one phase to the other. This implies that:

$$dG = \mu_1 dN_1 + \mu_2 dN_2 = 0, \qquad \mu_1(T,P) = \mu_2(T,P),$$

With dN1 = $-dN_2$. The condition of coexistence ($dG = \mu_1 dN_1 + \mu_2 dN_2 = 0,\ \mu_1(T,P) = \mu_2(T,P),$), tells us that the two phases 1 and 2 must have identical chemical potentials.

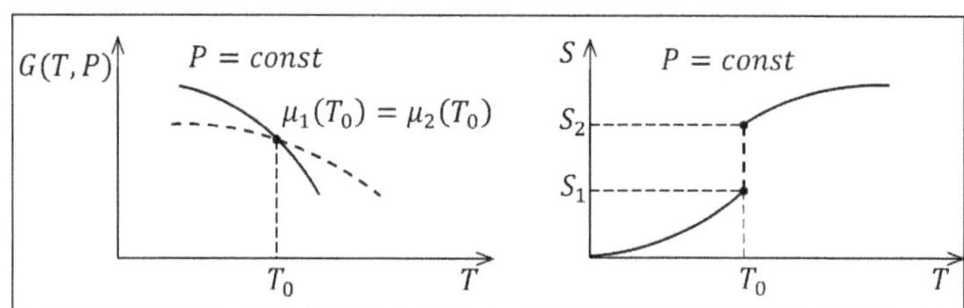

Clausius-Clapeyron Equation

We denote the discontinuities across the phase boundary with,

$$\Delta G = G_2(T,P,N_1) - G_1(T,P,N_2)$$

$$\Delta S = S_2 - S_1 = -\left(\frac{\partial(\Delta G)}{\partial T}\right)_p, \quad \Delta V = V_2 - V_1 = \left(\frac{\partial(\Delta G)}{\partial P}\right)_T,$$

Where we assume that $S_2 > S_1$.

Cyclic chain rule: The discontinuities ΔG, ΔS and ΔV are functions of V, T and P, which are in turn related by the equation of state f(P, V, T) = 0. There must hence exist a function $\tilde{f}$ such that:

$$\tilde{f}(\Delta G, T, P) = 0.$$

This condition allows to apply the cyclic chain rule. It leads to:

$$\underbrace{\left(\frac{\partial(\Delta G)}{\partial T}\right)_P}_{-\Delta S}\left(\frac{\partial T}{\partial P}\right)_{\Delta G}\underbrace{\left(\frac{\partial P}{\partial(\Delta G)}\right)_T}_{1/\Delta V} = -1,$$

where we have used $\Delta G = G_2(T,P,N_1) - G_1(T,P,N_2)$.

Vapor pressure: We note that P is the vapor pressure (Dampfdruck) for the gas-liquid transition. It changes with temperature along the phase transition line as,

$$\frac{dP}{dT} \equiv \left(\frac{\partial P}{\partial T}\right)_{\Delta G=0}$$

where we have used with ΔG = 0 when the two phases are in equilibrium with each other. $(\partial P/\partial T)_{\Delta G = 0}$ is hence slope of the transition line in the P–T diagram.

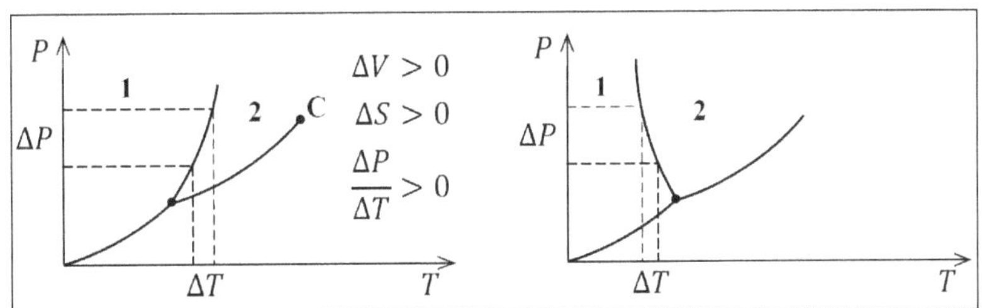

The slope ΔP/ΔT along the solid-liquid interface is positive when the substance contract upon freezing, the standard situation. Water does however expands upon freezing due to the Hydrogen bonding between molecules. ΔP/ΔT is in this case negative.

Clausius-Clapeyron relation: We rewrite $\underbrace{\left(\frac{\partial(\Delta G)}{\partial T}\right)_P}_{-\Delta S}\left(\frac{\partial T}{\partial P}\right)_{\Delta G}\underbrace{\left(\frac{\partial P}{\partial(\Delta G)}\right)_T}_{1/\Delta V} = -1$, as,

$$\left(\frac{dP}{dT}\right)_{\Delta G=0} = \frac{\Delta S}{\Delta V}, \quad \frac{dP}{dT} = \frac{\Delta Q_L}{T\Delta V},$$

Where ΔQ_L = T ΔS is the latent heat. All quantities entering the Clausius-Clapeyron relation can be measured. It can be used either as a consistency check or to determine the latent heat ΔQ by measuring T, ΔV and the slope dP/dT.

Second order transitions: In a second-order phase transition the first derivatives of G vanish and the Clapeyron equation is replaced by a condition involving second derivatives.

Ehrenfest Classification of Phase Transitions

Let us denote the two phases in equilibrium at a given coexistence curve as α and β. Following Ehrenfest we define next the order of the phase transition.

The order of the lowest derivative of the Gibbs enthalpy G showing a discontinuity upon crossing the coexistence curve is the order of a phase transition.

Phase transitions of order n. Explicitly, a phase transition between phases α and β is of order n if:

$$\left(\frac{\partial^m G_\alpha}{\partial T^m}\right)_P = \left(\frac{\partial^m G_\beta}{\partial T^m}\right)_P, \quad \left(\frac{\partial^m G_\alpha}{\partial P^m}\right)_T = \left(\frac{\partial^m G_\beta}{\partial P^m}\right)_T$$

For m = 1, 2, . . . , n − 1 and if:

$$\left(\frac{\partial^n G_\alpha}{\partial T^n}\right)_P \neq \left(\frac{\partial^n G_\beta}{\partial T^n}\right)_P, \quad \left(\frac{\partial^n G_\alpha}{\partial P^n}\right)_T \neq \left(\frac{\partial^n G_\beta}{\partial P^n}\right)_T$$

In practice, only phase transitions of first- and second-order are of importance. Their properties are listed below.

1st order:

- G(T, P) continuous.
- $S = -\left(\frac{\partial G}{\partial T}\right)_P$ *and* $V = \left(\frac{\partial G}{\partial P}\right)_T$ discontinuous;
- ∃ latent heat.

2nd order:

- G(T, P) continuous.
- S(T, P) and V (T, P) continuous.
- The discontinuities in the second order derivative of G(T, P, N) leads to discontinuities of the response functions:

$$C_P = T\left(\frac{\partial S}{\partial T}\right)_P = -T\left(\frac{\partial^2 G}{\partial T^2}\right)$$

$$k_T = -\frac{1}{V}\left(\frac{\partial V}{\partial P}\right)_T = -\frac{1}{V}\left(\frac{\partial^2 G}{\partial P^2}\right)_T$$

$$\alpha = \frac{1}{V}\left(\frac{\partial V}{\partial T}\right) = \frac{1}{V}\left(\frac{\partial^2 G}{\partial T \partial P}\right)$$

(susceptibilities) across the transition κ_T is the isothermal compressibility, α the thermal expansion coefficient and dG = −SdT + V dP + μdN.

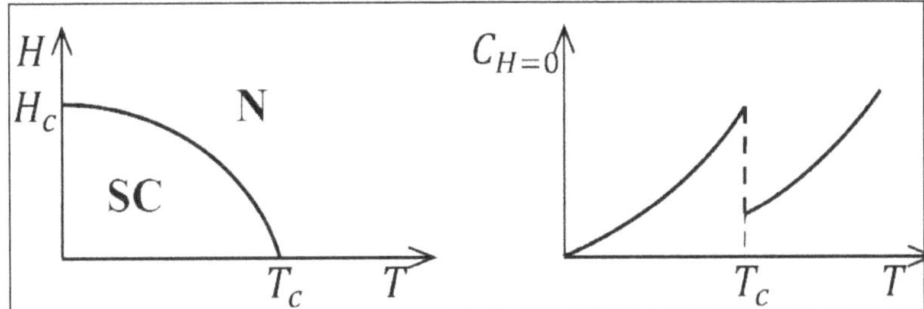

Specific Heat Jump: Phase transitions of the second order show a finite discontinuity in the specific heat C_p. An example is the transition to a superconducting state a at zero magnetic field $\mathcal{H}$.

Diverging Correlation Length at Criticality: The Ehrenfest classification is only valid if the motion of far away particles is not correlated, viz that the correlation length is finite. The correlation length diverges however at criticality for second order phase transition i.e. when $T \rightarrow T_c$. This leads in turn also to diverging response functions. The magnetic susceptibility $\mathcal{X}$ of a magnetic system diverges f. i. as:

$$\mathcal{X} = \left(\frac{\partial M}{\partial H}\right)_T, \quad \mathcal{X}(T) \sim \frac{1}{(T-T_c)^\gamma},$$

where γ is the critical exponent. Critical exponents are evaluated using advanced statistical mechanics methods, such as the renormalization group theory. Note, however, that $\mathcal{X} = \left(\frac{\partial M}{\partial H}\right)_T, \quad \mathcal{X}(T) \sim \frac{1}{(T-T_c)^\gamma}$, is observed in most cases only very close to the transition.

The entropy for a discontinuous transition: Discounting the exact order of transition we may classify a phase transition in any case with regard to the continuity of the entropy. For a discontinuous transition we have:

- $\Delta S \neq 0$; $\exists$latent heat;
- $C_p = -T\left(\frac{\partial^2 G}{\partial T^2}\right)_P$ is finite for $T \neq T_0$; no condition exists for $T = T_0$.

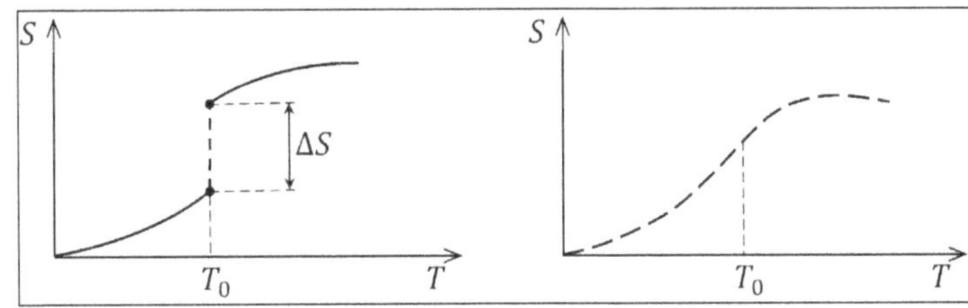

The entropy for a continuous transition. In this case we find:

- S continuous $\Rightarrow$ no latent heat.
- $\exists$ critical point T_c.
- Singularities in $C_V, k_T, \mathcal{X}_T$.

Density jumps: Consider a fluid system for which the volume V = (∂G/∂P)T shows a finite discontinuity ΔV at a 1st order phase transition, such as the liquid-gas line below the critical point. The corresponding densities particle density ρ = N/V is then likewise discontinuous.

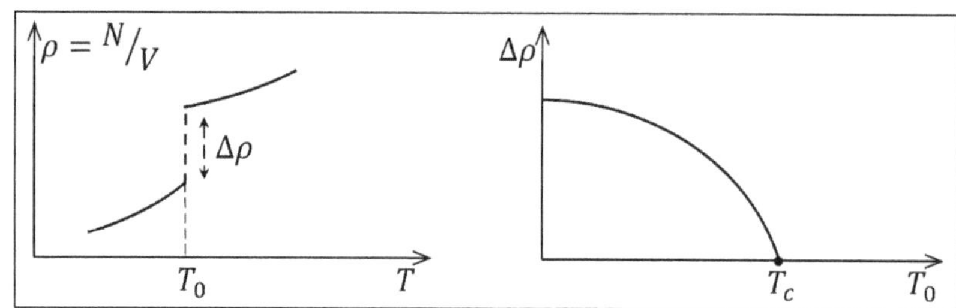

The density jump Δρ diminishes along the liquid-gas transition line, when the temperature is increased, until it vanishing at T_c. The transition becomes continuous at the critical point $T = T_c$.

Spontaneous magnetic ordering: The magnetization M of a magnetic compound is in part induced by an external magnetic field and in part due to the spontaneous ordering, viz by the alignment of the microscopic moments.

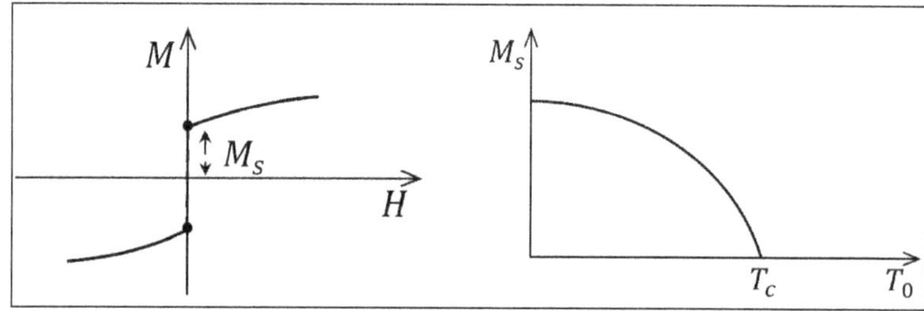

The magnetic work entering the internal energy dU = δQ + δW is δW = $\mathcal{H}$ dM, as defined by. With the Gibbs enthalpy G(T, $\mathcal{H}$) being the (two-fold) Legendre transform of the internal energy U(S, M), we then have that:

$$M = -\left(\frac{\partial G}{\partial \mathcal{H}}\right)_T, \quad dG = -SdT - Md\mathcal{H}$$

The magnetization M is discontinuous when spontaneous ordering is present, i.e. when the transition is of 1st order. Following the transition line, by increasing the temperature T □ T_c, the jump $2M_s$ decreases until spontaneous ordering disappears and the phase transition becomes second-order at $T_o = T_c$.

Van der Waals Equation of State

The equation of state for an ideal gas,

$$PV = Nk_B T, \qquad PV = nRT,$$

is only valid for very small densities of particles and, therefore, cannot describe the gasliquid phase transition. This phase transition is due to intermolecular interactions. we will derive an equation of state for a gas which includes intermolecular interaction in a phenomenological way.

Renormalized ideal gas: We consider with,

$$P_{eff}V_{eff} = nRT$$

The ansatz that a gas of interacting molecules still obeys the ideal gas equation of state $PV = Nk_BT$, $PV = nRT$, , albeit with yet to determine effective variables, where the effective pressure and temperature, P_{eff} and T_{eff}, are are functions of the physical pressure P and temperature T respectively. One can regard $P_{eff}V_{eff} = nRT$ as a renormalized ideal gas equation of state.

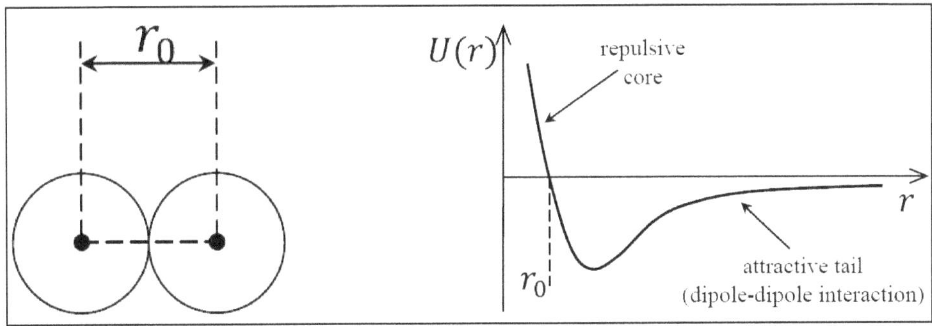

Interaction potential: The interaction between molecules is generically composed of a repulsive core (due to the Fermi repulsion between the electrons) and an attractive tail (due to the van der Waals dipole-dipole interaction).

The depth of the attractive U(r) is about 1 eV (changes with the gas species). This minimum is responsible for the chemical valence and for the crystal structure of solids.

Effective volume: The repulsive core of the intermolecular interaction leads to a volume exclusion, which can be modeled by considering the individual molecules as hard spheres of radius r_0. The effective Volume entering,

$$P_{eff}V_{eff} = nRT$$

is then,

$$V_{eff} = V - b'N, \qquad b' \approx \frac{1}{2}\frac{4\pi(2r_0)^3}{3} = \frac{16\pi r_0^3}{3},$$

where $N = n(R/k_B)$ is the overall number of molecules. Note that two hard-core particle of radius ro cannot come closer than $2r_0$, with the factor 1/2 in $V_{eff} = V - b'N,\ b' \approx \frac{1}{2}\frac{4\pi(2r_0)^3}{3} = \frac{16\pi r_0^3}{3}$, correcting for double counting.

Effective pressure: The interaction between molecules is pairwise and therefore proportional to the square of the density N/V. The van der Waals interaction mediated by induced dipoles in furthermore attractive. We may hence assume with,

$$P = P_{eff} - a'\frac{N^2}{V^2}, \qquad a' > 0$$

That the physical pressure P is smaller than the effective pressure P_{eff} by an amount proportional to (N/V)2.

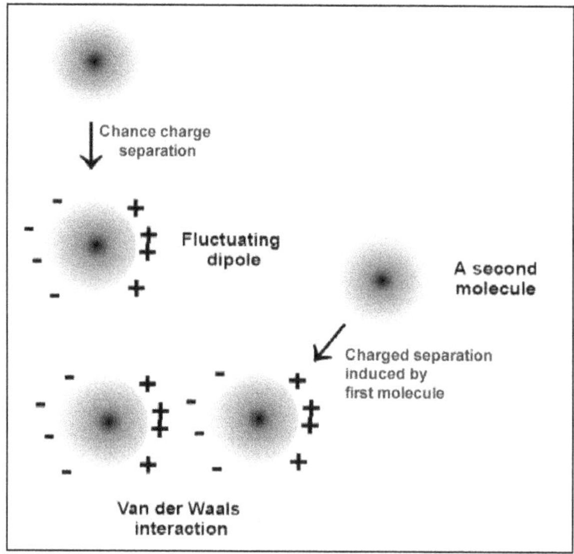

Van der Waals equation: We use,

$$a = N_a^2 a', \qquad b = N_a b', \qquad N = N_a n,$$

where NA is Avogadro's constant, together with the expressions,

$$V_{eff} = V - b'N, \qquad b' \approx \frac{1}{2}\frac{4\pi(2r_0)^3}{3} = \frac{16\pi r_0^3}{3},$$

and $P = P_{eff} - a'\frac{N^2}{V^2}$, $a' > 0$ for the effective volume V_{eff} = $V - N\,b'$ and respectively for the effective pressure P_{eff} = $P + a'N^2/V^2$. We then obtain the van der Waals equation,

$$P_{eff}V_{eff} = nRT, \qquad \left(P + a\left(\frac{n}{V}\right)^2\right)(V - bn) = nRT.$$

Virial Expansion

Inter-particle interactions become irrelevant in the low density limit (N/V) $\rightarrow$ 0, for which the the van der Waals equation $P_{eff}V_{eff} = nRT$, $\left(P + a\left(\frac{n}{V}\right)^2\right)(V - bn) = nRT.$), reduces consequently to the ideal-gas equation of state P V = nRT. It is hence of interest to evaluate the corrections to the ideal-gas equation of state by expanding $P_{eff}V_{eff} = nRT$, $\left(P + a\left(\frac{n}{V}\right)^2\right)(V - bn) = nRT.$ for fixed particle number N systematically in 1/V.

Rescaling: We start by rescaling the parameters of the van der Waals equation:

$$an^2 = A, \quad B = bn, \quad \overline{R} = nR.$$

We then obtain,

$$P = \frac{\overline{R}T}{(V-B)} - \frac{A}{V^2}, \quad \frac{PV}{\overline{R}T} = \left(1-\frac{B}{V}\right)^{-1} - \frac{A}{\overline{R}TV}.$$

For $P_{eff}V_{eff} = nRT$, $\left(P + a\left(\frac{n}{V}\right)^2\right)(V - bn) = nRT.$

Virial expansions: With the Taylor expansion,

$$\left(1-\frac{B}{V}\right)^{-1} = 1 + \frac{B}{V} + \left(\frac{B}{V}\right)^2 + \ldots$$

Of the first term in $P = \frac{\overline{R}T}{(V-B)} - \frac{A}{V^2}, \quad \frac{PV}{\overline{R}T} = \left(1-\frac{B}{V}\right)^{-1} - \frac{A}{\overline{R}TV}$. with respect to 1/V we obtain:

$$\frac{PV}{\overline{R}T} = 1 + \frac{1}{V}\left(B - \frac{A}{\overline{B}T}\right) + \left(\frac{B}{V}\right)^2 + \left(\frac{B}{V}\right)^3 + \ldots.$$

This expression has the form of a virial expansion:

$$\frac{PV}{\overline{R}T} = 1 + \frac{C_2}{V} + \frac{C_3}{V^2} + \ldots., \qquad C_2 = B - \frac{A}{\overline{R}T},$$

Where C_n is the nth virial coefficient. Measuring C_2 and $C_3 = B^2$ by observing experimentally the deviations from the ideal gas law on can extract the parameters A and B of the van der Waals gas. The virial expansion is also important for microscopic calculations.

Critical Point

The the isotherms of the pressure P are uniquely defined via the van der Waals equation $P = \frac{\overline{R}T}{(V-B)} - \frac{A}{V^2}, \quad \frac{PV}{\overline{R}T} = \left(1-\frac{B}{V}\right)^{-1} - \frac{A}{\overline{R}TV}$ by the Volume P. The reverse is however not true.

Rewriting $P_{eff}V_{eff} = nRT$, $\left(P + a\left(\frac{n}{V}\right)^2\right)(V - bn) = nRT.$

we obtain:

$$\left(PV^2 + an^2\right)(V - bn) = nRTV^2$$

Which shows that V (P) is given by the roots of a 3rd-order polynomial.

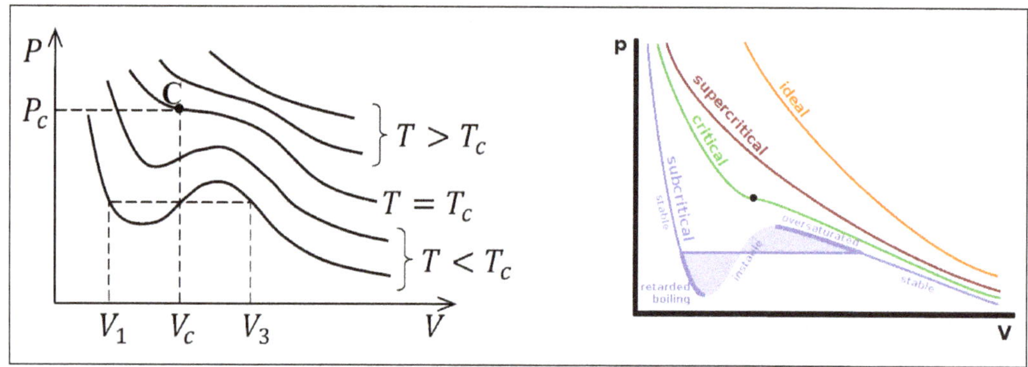

Roots of the Van der Waals equation: For given T and P there exist either three real roots or one real and two complex roots of $(PV^2 + an^2)(V - bn) = nRTV^2$. There exists hence a critical point (P_c, V_c, T_c) such that $(PV^2 + an^2)(V - bn) = nRTV^2$ has:

$$
\begin{aligned}
&T < T_c \;\; P < P_c \quad \text{three different real roots}\\
&T = T_c \;\; P = P_c \quad \text{one three-fold degenerate real root}\\
&T > T_c \;\; P > P_c \quad \text{one real root}
\end{aligned}
$$

The van der Waals of equation of the state must be then proportional to:

$$(V - V_c)^3 = 0, \qquad V^3 - 3V_cV^2 + 3V_c^2V - V_c^3 = 0$$

At the critical point (P_c, T_c, V_c). Comparing:

$$(PV^2 + an^2)(V - bn) = nRTV^2$$

with,

$$(V - V_c)^3 = 0, \qquad V^3 - 3V_cV^2 + 3V_c^2V - V_c^3 = 0$$

we find $V_c, P_c,$ and T_c :

$$
\left.
\begin{aligned}
3V_c &= nb + \frac{nRT_c}{Pc}\\
3V_c^2 &= \frac{an^2}{Pc}\\
V_c^3 &= \frac{abn^3}{Pc}
\end{aligned}
\right\}
\left\{
\begin{aligned}
V_c &= 3bn\\
P_c &= \frac{a}{27b^2}\\
RT_c &= \frac{8a}{27b}
\end{aligned}
\right.
$$

Comparison with the ideal gas: Combining the roots found in above question we can define with,

$$Z_c = \frac{P_cV_c}{nRT_c} = \frac{3}{8} = 0.375$$

A universal parameter measuring the deviation of a real gas from the ideal gas limit $Z_c \to 1$. For water we have $Z_c = 0.226$ and $T_c = 324$ °C. Real gases are generically further away from the ideal gas limit than the van der Waals theory would predict.

Maxwell Construction

The van der Waals isotherm is a monotonic function of V for $T > T_c$. For $T < T_c$, it has a "kink" with negative compressibility:

$$k_T = -\frac{1}{V}\left(\frac{\partial V}{\partial P}\right)_T < 0, \quad \left(\frac{\partial P}{\partial V}\right)_T > 0.$$

Negative compressibility, if existent for a real gas, would however lead to a collapse of the system, with a decreasing volume V and pressure P inducing each other. The system would not be thermodynamically stable.

Phase separation: The working substance avoids the collapse by separating spontaneously into two phases, with each of the two phases being located in a point in the V – P state phase characterized by a positive compressibility κ_T. These two phase will hence have different densities, corresponding respectively to a liquid and to a gas phase.

Coexistence condition: The two phases 1 and 2 are in equilibrium at the vapor pressure P_V, if the differentitials of the Gibbs enthalpies coincide for contant $N_1 + N_2 = N$ and $V_1 + V_2 = V$:

$$dG_1(T, P_V, N_1) = dG_2(T, P_V, N_2),$$
$$dF_1(T, V_1, N_1) + P_V dV_1 = dF_2(T, V_2, N_2) + P_V dV_2,$$

where we used that G(T, P, N) = F(T, V, N) + P V . Note that both the temperature T and the vapor pressure P_V are constant in $dF_1(T, V_1, N_1) + P_V dV_1 = dF_2(T, V_2, N_2) + P_V dV_2,$.

Free energy: The free energy F(T, V, N),

$$F(V,T) = -\int_{isotherm} P dV, \quad P = -\left(\frac{\partial F}{\partial V}\right)_{T,N},$$

Can be obtained as the area under the isotherm.

Integrating the coexistence condition $dF_1(T, V_1, N_1) + P_V dV_1 = dF_2(T, V_2, N_2) + P_V dV_2$, one obtains

$$F_1 - F_2 = P_V(V_2 - V_1),$$

Which implies that the volumes V_1 and V_2 are defined by a double tangent construction.

The free energy is is a weighted mixture of two phases 1 and 2 at any point along the tangent between 1 and 2. The resulting non-uniform state has the same P and T as the uniform state 3, but a lower free energy.

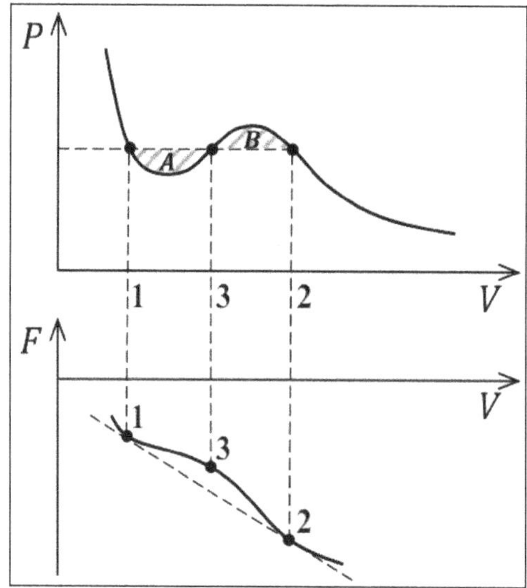

Maxwell construction: We note that,

$$F_2 - F_1 = \int_{V_1}^{V_2} (-P)\,dV = \int_{V_1}^{V_3} (-P)\,dV + \int_{V_3}^{V_2} (-P)\,dV$$

Along any isotherm, according to $F(V,T) = -\int_{isotherm} PdV,\ P = -\left(\frac{\partial F}{\partial V}\right)_{T,N}$. Rewriting the coexistence condition $F_2 - F_1 = \int_{V_1}^{V_2} (-P)\,dV = \int_{V_1}^{V_3} (-P)\,dV + \int_{V_3}^{V_2} (-P)\,dV$ as:

$$F_2 - F_1 = P_V (V_1 - V_3 + V_3 - V_2)$$

We then obtain with:

$$\underbrace{P_V (V_3 - V_1) - \int_{V_1}^{V_3} PdV}_{area\ A} = \underbrace{\int_{V_3}^{V_2} PdV - P_V (V_2 - V_3)}_{area\ B}$$

The Maxwell construction equation, $\underbrace{P_V (V_3 - V_1) - \int_{V_1}^{V_3} PdV}_{area\ A} = \underbrace{\int_{V_3}^{V_2} PdV - P_V (V_2 - V_3)}_{area\ B}$ determines the Vapor pressure $P_V = P(V_3)$ as the pressure for which the areas A and B are equal to each other.

References

- Entropy: sfu.ca, Retrieved 16 January, 2019
- The-Four-Laws-of-Thermodynamics/Second-Law-of-Thermodynamics, Supplemental-Modules-(Physical-and-Theoretical-Chemistry): chem.libretexts.org, Retrieved 15 May, 2019
- What-is-clausius-theorem: hkdivedi.com, Retrieved 05 April, 2019
- Clausius-inequality: sciencedirect.com, Retrieved 21 June, 2019
- Phase-transitions: itp.uni-frankfurt.de, Retrieved 27 July, 2019

5

Thermodynamic Potentials

Thermodynamic potentials deals with the representation of the thermodynamic state of a system. It includes internal energy, enthalpy, Gibbs Free Energy and the Helmholtz Free Energy. This chapter has been carefully written to provide an easy understanding of these thermodynamic potentials.

A thermodynamic potential is a scalar quantity. It represents the thermodynamic state of a system. The concept of thermodynamic potentials was introduced by Pierre Duhem in1886. Josiah Willard Gibbs called these as fundamental functions. These potentials have been derived from the concept of potential energy. It is well known that potential energy is the capacity to do work. Similarly internal energy (U) is the capacity to do useful work plus the capacity to release heat energy. Enthalpy (H) is the capacity to do non- mechanical work plus the heat energy given to the system. Non mechanical work means like chemical energy/electrical energy/thermal energy/acoustic energy/nuclear energy/radiant energy. These are not due to position or motion of a body i.e. these are other forms of energy. Helmholtz energy (F) is the capacity to do reversible work or useful work (mechanical work). Gibbs energy (G) is the capacity to do non- mechanical work. From the above, it can be easily concluded that if ΔU is the energy added to a system, ΔF is the total work done, ΔG is the non- mechanical work done while ΔH is the sum of non- mechanical work done and the heat given to it. Thermodynamic potentials are useful in the thermodynamics of chemical reactions. These potentials also describe the non-cyclic processes. They are four thermodynamic potentials as mentioned in table. These are internal energy, the enthalpy, the Helmholtz free energy and the Gibbs free energy. All the thermodynamic potentials are energy terms. These potentials cannot be measured like temperature or pressure. These can be found from the measurable variables like pressure, volume and temperature. Thermodynamic potentials, fundamental equation and equilibrium conditions involve the terms +p V and –TS. All the thermodynamic properties of a system can be found by taking the partial derivatives of the thermodynamic potentials as given below:

$$(\partial U / \partial T)V = T$$

$$(\partial F / \partial P)T = -P$$

$$(\partial G / \partial V)T = V$$

Further ΔF is the measure of useful work under constant temperature (T) and constant volume (V). ΔG is the useful work done under constant temperature (T) and constant pressure (P).

Table: Thermodynamic potentials and formulas.

S. No.	Name	Symbol	Formula
1.	Internal Energy	U	Q – pV
2.	Enthalpy	H	U + pV
3.	Helmholtz free energy	F	U – TS
4.	Gibbs Free Energy	G	U + pV – TS

Table: Thermodynamic potentials and natural variables.

S. No.	Thermodynamic potential	Natural variables (which are kept constant in the process)
1.	U	S, V
2.	H	S, p
3.	G	T, p
4.	F	T, V

Table: Thermodynamic potentials, fundamentale quation and equilibrium condition.

S. No.	Thermodynamic potential	Fundamental equation	Equilibrium condition
1	S (U, V, N)	Entropy fundamental equation	Maximum
2	U (S, V, N)	Internal Energy fundamental equation	Maximum
3	H (S, P,N)	Enthalpy fundamental equation	Minimum
4	F (T, V, N)	Helmsholtz free energy fundamental equation	Minimum
5	G (T, P, N)	Gibbs free energy fundamental equation	Minimum

N = Number of particles related to molecular weight.

These thermodynamic potentials are connected with experimentally measureable thermodynamic parameters and thus make their estimations indirectly. It is noticeable that thermodynamic potentials have multiple independent variables. There is an important feature of all the multiple variable functions that the mixed partial derivatives do not depend on the order in which these are used like $\partial 2f/\partial x\,\partial y = \partial 2f/\partial y\,\partial x$ (It is Euler' Reciprocity Law of thermodynamics).

Natural Variables

The variables which are kept constant in a process are called the natural variables of that potential. Thermodynamic potential can be found as a function of its natural variables only. Then only the thermodynamic properties of the system can be determined by taking partial derivatives of that potential with respect to its natural variables. If the thermodynamic potential is not known in terms of its natural variables, then thermodynamic properties cannot be found from the partial derivatives of that potential.

Natural variables for the four thermodynamic potentials are formed from every combination of the T-S and P-V variables, excluding any pairs of conjugate variables. The conjugate pairs are made with μi and Ni.

Where μi is the chemical potential for an i-type particle. Ni is the number of particles of type i in the system.

Internal Energy

In thermodynamics, a *system* is any object, any quantity of matter, any region, etc. selected for study and mentally set apart from everything else which is then called its surroundings. The imaginary envelope enclosing the system and separating it from its surroundings is called the boundary of the system. The boundaries will be referred to as the *walls* of the system.

The internal energy of a system is simply its energy. The term was introduced into thermodynamics in 1852 by W. Thomson (the later Lord Kelvin). The adjective "internal" refers to the fact that some energy contributions are not considered. For instance, when the total system is in uniform motion, it has kinetic energy. This overall kinetic energy is never seen as part of the internal energy; one could call it *external energy*. Or, if the system is at constant non-zero height above the surface the Earth, it has constant potential energy in the gravitational field of the Earth. Gravitational energy is only taken into account when it plays a role in the phenomenon of interest, for instance in a colloidal suspension, where the gravitation influences the up- downward motion of the small particles comprising the colloid. In all other cases, gravitational energy is assumed not to contribute to the internal energy; one may call it again external energy.

On the other hand, a contribution to internal energy that is *always* included is the kinetic energy of the atoms or molecules constituting the system. In an atomic gas, it is the energy associated with translations of the atoms; in a molecular gas translations and molecular rotations contribute to internal energy. In a solid, internal energy acquires contributions from vibrations, among other effects. Except for ideal gases, the potential energy of molecules in the field of the others is also an important component of the internal energy.

In general, energies that are not changing in the processes of interest are left out of the definition of internal energy. For instance, when a system consists of a vessel filled with water and the process of interest is evaporation (formation of steam), the kinetic energy of the water molecules and the interaction between them are included in the internal energy. As long as no chemical bonds are broken, the energies contained in these bonds are not included. If the temperatures are not too high, say below 200 to 300 °C, the intramolecular vibrational energies are ignored as well. Chemists and engineers never include relativistic contributions, of the type $E = mc^2$, or nuclear contributions (say the fusion energy of protons with oxygen-nuclei). However, a plasma physicist studying the thermodynamics of fusion reactions will include nuclear energy in the internal energy of the plasma.

First Law of Thermodynamics and Internal Energy

Classical (phenomenological) thermodynamics is not concerned with the nature of the internal energy, it simply postulates that it exists and may be changed by certain processes. Further it is postulated that internal energy, usually denoted by either U or E, is a state function, that is, its value depends upon the state of the system and not upon the nature or history of the past processes by which the system attained its state. In addition, the internal energy, which henceforth will be written as U, is assumed to be a differentiable function of the independent variables that uniquely specify the state of the system. An example of such a state variable is the volume V of the system.

When the system has thermally conducting walls, an amount of heat DQ can go through the wall in either direction: if $DQ > 0$, heat enters the system and if $DQ < 0$ the system loses heat to its surroundings. The symbol DQ indicates simply a small amount of heat, and not a differential of Q. Note that, because it is not a function, Q does not have a differential. The internal energy of the system changes by dU as a consequence of the heat flow, and it is postulated that,

$$dU = DQ$$

Reiterating, the usual sign convention is such that positive DQ is the heat absorbed by the system, i.e., the heat that the system receives from its surroundings. The symbol dU indicates a differential of the differentiable function U.

Most thermodynamic systems are such that work can be performed *on* them or *by* them. When a small amount of work DW is performed *by* the system, the internal energy decreases,

$$dU = -DW$$

The sign convention is such that DW is the work *by* the system on its surroundings, hence the minus sign in this equation.

As an example of work, consider a cylinder of volume V that may be changed by moving a piston in or out. The cylinder contains gas of pressure p. A small amount of work pdV is performed *on* the system by reversibly (quasi-statically) moving the piston inward ($dV < 0$). The sign convention of DW is such that DW and dV have the same sign,

$$DW = pdV \quad dV < 0, \quad DW < 0$$

Due to the fact that the work is performed reversibly, the small amount of work DW is proportional to the differential dV. If dV > 0 (expansion), work DW > 0 is performed by the system. Hence the change in internal energy obtains indeed a minus sign:

$$dU = -DW = -pdV$$

Note that other forms of work than pdV are possible. For instance, $DW = -HdM$, the product of an external magnetic field H with a small change in total magnetization dM, is a change in internal energy caused by an alignment of the microscopic magnetic moments that constitute a magnetizable material.

An important form of doing work is the reversible addition of substance,

$$DW = -\mu dn,$$

Here μ (a function of thermodynamic parameters as T, p, etc.) is the chemical potential of the pure substance added to the system. The infinitesimal quantity dn is the amount (expressed in moles) of substance added. The chemical potential μ is the amount of energy that the system gains when reversibly, adiabatically ($DQ = 0$), and isochorically ($dV = 0$) a mole of substance is added to it.

When a small amount of heat DQ flows in or out the system and simultaneously a small amount of work DW is done by or on the system, the first law of thermodynamics states that the internal energy changes as follows:

$$dU = DQ - DW$$

Note that the sum of two small quantities, both not necessarily differentials, gives a differential of the state function U. The first law, equation $dU = DQ - DW$, postulates the existence of a state function that accumulates the work done on/by the system and the heat that flows in/out the system.

Internal energy is an extensive property—that is, its magnitude depends on the amount of substance in a given state. Often one considers the *molar energy*, energy per amount of substance (amount expressed in moles); this is an intensive property. Also the *specific energy* (energy per kilogram) is an intensive property. The (extensive) internal energy has the SI dimension joule.

Note that thus far only a *change* in internal energy was defined. An absolute value can be obtained by defining a zero (reference) point with $U_0 = 0$ and integration,

$$\int_0^1 dU = U_1 - U_0 = U_1$$

Since U is a state function U_1 is independent of the integration path (the choice of values of S, V, and n between lower and upper bound of the integration). The reference point U_0 could be at the zero of absolute temperature (zero kelvin).

Explicit Expression

Consider a one-component thermodynamical system that allows heat exchange DQ, work $-pdV$, and matter exchange μdn. The second law of thermodynamics states that there exists a variable, entropy (commonly denoted by S) that is given by:

$$dS = \frac{DQ}{T},$$

That is, the integrating factor $1/T$ converts the small quantity DQ into the differential dS. This relation holds when the heat exchange occurs reversibly. By the second law, the entropy S is a state variable. It is size-extensive, i.e., S is linear in the size of the system, and has dimension J/K (joule per degree kelvin).

Since there are three forms of energy contact with the surroundings, the system has exactly three independent state variables. Choose now as independent set the variables S, V, and n, which all three are size-extensive. By results shown earlier, cf. equation $dU = DQ - DW$, and use of $DQ = TdS$, the differential of the internal energy is,

$$dU = TdS - pdV + \mu dn.$$

It will be shown that an explicit expression for the internal energy U of the system under consideration is,

$$U = TS - pV + \mu n.$$

In order to prove $U = TS - pV + \mu n.$, we consider first two identical systems with the same values for the three size-extensive independent variables *S, V, n* (and hence also the same values for *U, T,* and *p*). Clearly, the "supersystem" consisting of the two identical systems has twice the entropy, volume, and amount of substance, and the sum of the energies of the two systems is 2*U*, or

$$U(2S, 2V, 2n) = 2U(S, V, n)$$

The same kind of equation holds when we separate the original system into two equal parts, then the energy, entropy, volume and amount of substance are halved for each of the two parts. Clearly then, for an arbitrary real positive number λ,

$$U(\lambda S, \lambda V, \lambda n) = \lambda U(S, V, n)$$

That is, the internal energy U is a homogeneous function of order 1 of the size-extensive variables, S, V and n. By Euler's theorem,

$$U = \left(\frac{\partial U}{\partial S}\right)_{V,n} S + \left(\frac{\partial U}{\partial V}\right)_{S,n} V + \left(\frac{\partial U}{\partial n}\right)_{S,V} n.$$

Since U is a function of the three variables, dU can be written in terms of partial derivatives. When this is equated to the expression in equation $dU = TdS - pdV + \mu dn.$

$$\begin{aligned} dU &= \left(\frac{\partial U}{\partial S}\right)_{V,n} dS + \left(\frac{\partial U}{\partial V}\right)_{S,n} dV + \left(\frac{\partial U}{\partial n}\right)_{S,V} dn \\ &= TdS - pdV + \mu dn \end{aligned}$$

it follows (since S, V, and n are independent) that,

$$T = \left(\frac{\partial U}{\partial S}\right)_{V,n}, \quad p = -\left(\frac{\partial U}{\partial V}\right)_{S,n}, \quad \mu = \left(\frac{\partial U}{\partial n}\right)_{S,V}$$

Insertion of this into equation $U = \left(\frac{\partial U}{\partial S}\right)_{V,n} S + \left(\frac{\partial U}{\partial V}\right)_{S,n} V + \left(\frac{\partial U}{\partial n}\right)_{S,V} n.$

gives the result $U = TS - pV + \mu n.$.

Statistical Thermodynamics Definition

Consider a system of constant temperature *T*, constant number of molecules *N*, and constant volume *V*. In statistical thermodynamics one defines for such a system the density operator.

$$\hat{\rho} \stackrel{\text{def}}{=} \frac{e^{-\beta \hat{H}}}{\mathrm{Tr}(e^{-\beta \hat{H}})} = \frac{e^{-\beta \hat{H}}}{Q} \quad \text{with} \quad Q \stackrel{\text{def}}{=} \mathrm{Tr}(e^{-\beta \hat{H}}),$$

where $\hat{H}$ is the Hamiltonian (energy operator) of the total system, $\mathrm{Tr}(\hat{O})$ is the trace of the operator $\hat{O}$, $\beta = 1/(kT)$, and k is Boltzmann's constant. The quantity Q is the partition function.

The thermodynamic average of $\hat{H}$ is equal to the internal energy U,

$$U = \langle\langle \hat{H} \rangle\rangle \overset{\text{def}}{=} \mathrm{Tr}(\hat{\rho}\,\hat{H}) = \frac{1}{Q}\mathrm{Tr}(\hat{H}\,e^{-\beta\hat{H}})$$

The internal energy is minus the logarithmic derivative of Q,

$$\frac{d\ln Q}{d\beta} = \frac{1}{Q}\frac{dQ}{d\beta} = \frac{1}{Q}\mathrm{Tr}\frac{d(e^{-\beta\hat{H}})}{d\beta} = -\frac{1}{Q}\mathrm{Tr}(\hat{H}\,e^{-\beta\hat{H}}) = -U$$

Further

$$\frac{d\ln Q}{d\beta} = \frac{d\ln Q}{dT}\left(\frac{d\beta}{dT}\right)^{-1} = -kT^2\,\frac{d\ln Q}{dT}.$$

Hence, the following well-known statistical-thermodynamics expression is obtained for the internal energy U,

$$U = kT^2\left(\frac{d\ln Q}{dT}\right) = kT^2\left(\frac{\partial \ln Q}{\partial T}\right)_{V,N}$$

- The existence of an energy operator $\hat{H}$ was simply assumed. The choice of energy terms to be included in this operator, is in fact equivalent to the choice of contributions adding to the internal energy *U*. Statistical thermodynamics does not solve the problem of defining internal energy.
- When the trace is evaluated in a basis of eigenstates of $\hat{H}$, the physical meaning of the density operator becomes clearer. In fact, Boltzmann weight factors will arise. Thus, upon writing,

$$\hat{H}\,|E_i\rangle = E_i\,|E_i\rangle, \qquad e^{-\beta\hat{H}}\,|E_i\rangle = e^{-\beta E_i}\,|E_i\rangle, \qquad \mathrm{Tr}(e^{-\beta\hat{H}}) = \sum_i \langle E_i|e^{-\beta\hat{H}}\,|E_i\rangle,$$

The partition function becomes [$\beta = 1/(kT)$]:

$$Q = \sum_i e^{-E_i/(kT)}, \qquad \text{(sum over eigenstates, not over energy levels)},$$

And the thermodynamic average becomes in a basis of energy eigenstates,

$$U = \langle\langle \hat{H} \rangle\rangle = \frac{1}{Q}\sum_i E_i\,e^{-E_i/(kT)}.$$

The partition function normalizes the Boltzmann weights, $\exp[-E_i/(kT)]$. Indeed,

$$\mathrm{Tr}(\hat{\rho}) = \frac{1}{Q}\left[\sum_i e^{-E_i/(kT)}\right] = \frac{Q}{Q} = 1.$$

The sum over normalized weights equals unity, as a proper weight function should.

Gibbs Free Energy

The Gibbs free energy (G), often called simply free energy, was named in honor of J. Willard Gibbs (1838–1903), an American physicist who first developed the concept. It is defined in terms of three other state functions with which you are already familiar: enthalpy, temperature, and entropy:

$$G = H - TS$$

Because it is a combination of state functions, G is also a state function.

The criterion for predicting spontaneity is based on (ΔG), the change in G, at constant temperature and pressure. Although very few chemical reactions actually occur under conditions of constant temperature and pressure, most systems can be brought back to the initial temperature and pressure without significantly affecting the value of thermodynamic state functions such as G. At constant temperature and pressure,

$$\Delta G = \Delta H - T\Delta S$$

where all thermodynamic quantities are those of the system. Recall that at constant pressure, ΔH = q , whether a process is reversible or irreversible, and $T\Delta S = q_{rev}$. Using these expressions, we can reduce Equation $\Delta G = \Delta H - T\Delta S$ to $\Delta G = q - q_{rev}$. Thus ΔG is the difference between the heat released during a process (via a reversible or an irreversible path) and the heat released for the same process occurring in a reversible manner. Under the special condition in which a process occurs reversibly, $q = q_{rev}$ and ΔG = 0. if ΔG is zero, the system is at equilibrium, and there will be no net change.

What about processes for which ΔG ≠ 0? To understand how the sign of ΔG for a system determines the direction in which change is spontaneous, we can rewrite the relationship between ΔS and q_{rev},.

$$\Delta S = \frac{q_{rev}}{T}$$

With the definition of ΔH in terms of q_{rev},

$$q_{rev} = \Delta H$$

To obtain,

$$\Delta S_{surr} = -\frac{\Delta H_{sys}}{T}$$

Thus the entropy change of the surroundings is related to the enthalpy change of the system. We have stated that for a spontaneous reaction, $\Delta S_{univ} > 0$, so substituting we obtain,

$$\Delta S_{univ} = \Delta S_{sys} + \Delta S_{surr} > 0$$

$$= \Delta S_{sys} - \frac{\Delta H_{sys}}{T} > 0$$

Multiplying both sides of the inequality by –T reverses the sign of the inequality; rearranging,

$$\Delta H_{sys} - T\Delta S_{sys} < 0$$

Which is equal to ΔG. We can therefore see that for a spontaneous process, ΔG < 0.

The relationship between the entropy change of the surroundings and the heat gained or lost by the system provides the key connection between the thermodynamic properties of the system and the change in entropy of the universe. The relationship shown in Equation $\Delta G = \Delta H - T\Delta S$ allows us to predict spontaneity by focusing exclusively on the thermodynamic properties and temperature of the system. We predict that highly exothermic processes (ΔH < < 0) that increase the disorder of a system (ΔS_{sys} > > 0) would therefore occur spontaneously. An example of such a process is the decomposition of ammonium nitrate fertilizer. Ammonium nitrate was also used to destroy the Murrah Federal Building in Oklahoma City, Oklahoma, in 1995. For a system at constant temperature and pressure, we can summarize the following results:

- If ΔG < 0, the process occurs spontaneously.
- If ΔG = 0, the system is at equilibrium.
- If ΔG > 0, the process is not spontaneous as written but occurs spontaneously in the reverse direction.

To further understand how the various components of ΔG dictate whether a process occurs spontaneously, we now look at a simple and familiar physical change: the conversion of liquid water to water vapor. If this process is carried out at 1 atm and the normal boiling point of 100.00°C (373.15 K), we can calculate ΔG from the experimentally measured value of ΔH_{vap} (40.657 kJ/mol). For vaporizing 1 mol of water, ΔH = 40,657; J, so the process is highly endothermic. From the definition of ΔS, we know that for 1 mol of water,

$$\Delta S_{vap} = \frac{\Delta H_{vap}}{T_b} = \frac{10,657\ J}{373.15\ K} = 108.96\ J/K$$

Hence there is an increase in the disorder of the system. At the normal boiling point of water,

$$\Delta G_{100°C} = \Delta H_{100°C} - T\Delta S_{100°C} = 40,657\ J - [(373.15\ K)(108.96\ J/K)] = 0\ J$$

The energy required for vaporization offsets the increase in disorder of the system. Thus ΔG = 0, and the liquid and vapor are in equilibrium, as is true of any liquid at its boiling point under standard conditions.

Now suppose we were to superheat 1 mol of liquid water to 110°C. The value of ΔG for the vaporization of 1 mol of water at 110°C, assuming that ΔH and ΔS do not change significantly with temperature, becomes,

$$\Delta G_{110°C} = \Delta H - T\Delta S = 40,657\ J - [(383.15\ K)(108.96\ J/K)] = -1091\ J$$

At 110°C, ΔG < 0, and vaporization is predicted to occur spontaneously and irreversibly.

We can also calculate ΔG for the vaporization of 1 mol of water at a temperature below its normal boiling point—for example, 90°C—making the same assumptions:

$$\Delta G_{90°C} = \Delta H - T\Delta S = 40,657 \text{ J} - [(363.15 \text{ K})(108.96 \text{ J/K})] = 1088 \text{ J}$$

At 90°C, ΔG > 0, and water does not spontaneously convert to water vapor. When using all the digits in the calculator display in carrying out our calculations, $\Delta G_{110°C}$ = 1090 J = $-\Delta G_{90°C}$, as we would predict.

Equilibrium Conditions

ΔG = 0 only if ΔH = TΔS.

We can also calculate the temperature at which liquid water is in equilibrium with water vapor. Inserting the values of ΔH and ΔS into the definition of ΔG, setting ΔG = 0, and solving for T,

$$0 \text{ J} = 40,657 \text{ J} - T(108.96 \text{ J/K})$$
$$T = 373.15 \text{ K}$$

Thus ΔG = 0 at T = 373.15 K and 1 atm, which indicates that liquid water and water vapor are in equilibrium; this temperature is called the normal boiling point of water. At temperatures greater than 373.15 K, ΔG is negative, and water evaporates spontaneously and irreversibly. Below 373.15 K, ΔG is positive, and water does not evaporate spontaneously. Instead, water vapor at a temperature less than 373.15 K and 1 atm will spontaneously and irreversibly condense to liquid water. Figure shows how the ΔH and TΔS terms vary with temperature for the vaporization of water. When the two lines cross, ΔG = 0, and ΔH = TΔS.

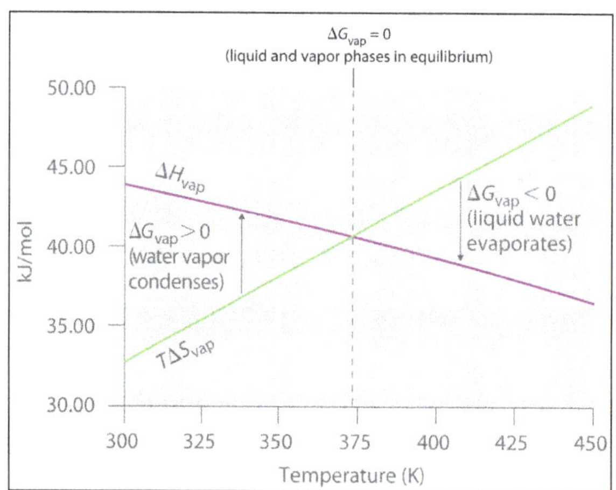

Temperature Dependence of ΔH and TΔS for the Vaporization of Water. Both ΔH and TΔS are temperature dependent, but the lines have opposite slopes and cross at 373.15 K at 1 atm, where ΔH = TΔS. Because ΔG = ΔH – TΔS, at this temperature ΔG = 0, indicating that the liquid and vapor phases are in equilibrium. The normal boiling point of water is therefore 373.15 K. Above the normal boiling point, the TΔS term is greater than ΔH, making ΔG < 0; hence, liquid water evaporates spontaneously. Below the normal boiling point, the ΔH term is greater than TΔS, making ΔG > 0. Thus liquid water does not evaporate spontaneously, but water vapor spontaneously condenses to liquid.

A similar situation arises in the conversion of liquid egg white to a solid when an egg is boiled. The major component of egg white is a protein called albumin, which is held in a compact, ordered structure by a large number of hydrogen bonds. Breaking them requires an input of energy ($\Delta H > 0$), which converts the albumin to a highly disordered structure in which the molecules aggregate as a disorganized solid ($\Delta S > 0$). At temperatures greater than 373 K, the $T\Delta S$ term dominates, and $\Delta G < 0$, so the conversion of a raw egg to a hard-boiled egg is an irreversible and spontaneous process above 373 K.

The Relationship between ΔG and Work

The value of ΔG allows us to predict the spontaneity of a physical or a chemical change. In addition, the magnitude of ΔG for a process provides other important information. The change in free energy (ΔG) is equal to the maximum amount of work that a system can perform on the surroundings while undergoing a spontaneous change (at constant temperature and pressure): $\Delta G = w_{max}$. To see why this is true, let's look again at the relationships among free energy, enthalpy, and entropy expressed in Equation $\Delta G = \Delta H - T\Delta S$. We can rearrange this equation as follows:

$$\Delta H = \Delta G + T\Delta S$$

This equation tells us that when energy is released during an exothermic process ($\Delta H < 0$), such as during the combustion of a fuel, some of that energy can be used to do work ($\Delta G < 0$), while some is used to increase the entropy of the universe ($T\Delta S > 0$). Only if the process occurs infinitely slowly in a perfectly reversible manner will the entropy of the universe be unchanged. Because no real system is perfectly reversible, the entropy of the universe increases during all processes that produce energy. As a result, no process that uses stored energy can ever be 100% efficient; that is, ΔH will never equal ΔG because ΔS has a positive value.

One of the major challenges facing engineers is to maximize the efficiency of converting stored energy to useful work or converting one form of energy to another. As indicated in table, the efficiencies of various energy-converting devices vary widely. For example, an internal combustion engine typically uses only 25%–30% of the energy stored in the hydrocarbon fuel to perform work; the rest of the stored energy is released in an unusable form as heat. In contrast, gas–electric hybrid engines, now used in several models of automobiles, deliver approximately 50% greater fuel efficiency. A large electrical generator is highly efficient (approximately 99%) in converting mechanical to electrical energy, but a typical incandescent light bulb is one of the least efficient devices known (only approximately 5% of the electrical energy is converted to light). In contrast, a mammalian liver cell is a relatively efficient machine and can use fuels such as glucose with an efficiency of 30%–50%.

Table: Approximate thermodynamic efficiencies of various devices.

Device	Energy Conversion	Approximate Efficiency (%)
large electrical generator	mechanical → electrical	99
chemical battery	chemical → electrical	90
home furnace	chemical → heat	65
small electric tool	electrical → mechanical	60

space shuttle engine	chemical → mechanical	50
mammalian liver cell	chemical → chemical	30–50
spinach cell	light → chemical	30
internal combustion engine	chemical → mechanical	25–30
fluorescent light	electrical → light	20
solar cell	light → electricity	10
incandescent light bulb	electricity → light	5
yeast cell	chemical → chemical	2–4

Standard Free-Energy Change

We have seen that there is no way to measure absolute enthalpies, although we can measure changes in enthalpy (ΔH) during a chemical reaction. Because enthalpy is one of the components of Gibbs free energy, we are consequently unable to measure absolute free energies; we can measure only changes in free energy. The standard free-energy change (ΔG°) is the change in free energy when one substance or a set of substances in their standard states is converted to one or more other substances, also in their standard states. The standard free-energy change can be calculated from the definition of free energy, if the standard enthalpy and entropy changes are known, using equation $\Delta G^{\circ} = \Delta H^{\circ} - T\Delta S^{\circ}$:

$$\Delta G^{\circ} = \Delta H^{\circ} - T\Delta S^{\circ}$$

If ΔS° and ΔH° for a reaction have the same sign, then the sign of ΔG° depends on the relative magnitudes of the ΔH° and TΔS° terms. It is important to recognize that a positive value of ΔG° for a reaction does not mean that no products will form if the reactants in their standard states are mixed; it means only that at equilibrium the concentrations of the products will be less than the concentrations of the reactants.

A positive ΔG° means that the equilibrium constant is less than 1.

Example:

Calculate the standard free-energy change (ΔG°) at 25 °C for the reaction:

$$H_{2(g)} + O_{2(g)} \rightleftharpoons H_2O_{2(l)}$$

At 25°C, the standard enthalpy change (ΔH°) is −187.78 kJ/mol, and the absolute entropies of the products and reactants are:

- S°(H_2O_2) = 109.6 J/(mol•K),
- S°(O_2) = 205.2 J/(mol•K), and
- S°(H_2) = 130.7 J/(mol•K).

Is the reaction spontaneous as written?

Given: balanced chemical equation, ΔH° and S° for reactants and products.

Asked for: spontaneity of reaction as written.

Strategy:

- Calculate $\Delta S°$ from the absolute molar entropy values given.
- Use Equation $\Delta G^{\circ} = \Delta H^{\circ} - T\Delta S^{\circ}$, the calculated value of $\Delta S°$, and other data given to calculate $\Delta G°$ for the reaction. Use the value of $\Delta G°$ to determine whether the reaction is spontaneous as written.

Solution:

A To calculate $\Delta G°$ for the reaction, we need to know $\Delta H°$, $\Delta S°$, and T. We are given $\Delta H°$, and we know that T = 298.15 K. We can calculate $\Delta S°$ from the absolute molar entropy values provided using the "products minus reactants" rule:

$$\begin{aligned} \Delta S° &= S°(H_2O_2) - [S°(O_2) + S°(H_2)] \\ &= [1\,\text{mol}\ H_2O_2 \times 109.6\,J/(\text{mol}\cdot\text{K})] \\ &= \{[1\,\text{mol}\ \text{H}_2 \times 130.7\,J/(\text{mol}\cdot K)] + [1\,\text{mol}\ \text{O}_2 \times 205.2\,J/(\text{mol}\cdot\text{K})]\} \\ &= -226.3\ J/K\,(\text{per mole of H}_2\text{O}_2) \end{aligned}$$

As we might expect for a reaction in which 2 mol of gas is converted to 1 mol of a much more ordered liquid, $\Delta S°$ is very negative for this reaction.

B Substituting the appropriate quantities into Equation $\Delta G^{\circ} = \Delta H^{\circ} - T\Delta S^{\circ}$,

$$\begin{aligned} \Delta G° &= \Delta H° - T\Delta S° = -187.78\,\text{kJ/mol} - (298.15\ \text{K})[-226.3\text{J}/(\text{mol}\cdot\text{K}) \times 1\ \text{kJ}/1000\ \text{J}] \\ &= -187.78\,\text{kJ/mol} + 67.47\,\text{kJ/mol} = -120.31\,\text{kJ/mol} \end{aligned}$$

The negative value of $\Delta G°$ indicates that the reaction is spontaneous as written. Because $\Delta S°$ and $\Delta H°$ for this reaction have the same sign, the sign of $\Delta G°$ depends on the relative magnitudes of the $\Delta H°$ and $T\Delta S°$ terms. In this particular case, the enthalpy term dominates, indicating that the strength of the bonds formed in the product more than compensates for the unfavorable $\Delta S°$ term and for the energy needed to break bonds in the reactants.

Tabulated values of standard free energies of formation allow chemists to calculate the values of $\Delta G°$ for a wide variety of chemical reactions rather than having to measure them in the laboratory. The standard free energy of formation (ΔG_f°) of a compound is the change in free energy that occurs when 1 mol of a substance in its standard state is formed from the component elements in their standard states. By definition, the standard free energy of formation of an element in its standard state is zero at 298.15 K. One mole of Cl_2 gas at 298.15 K, for example, has $\Delta G_f^{\circ} = 0$. The standard free energy of formation of a compound can be calculated from the standard enthalpy of formation (ΔH°_f) and the standard entropy of formation (ΔS°_f) using the definition of free energy:

$$\Delta_f^o = \Delta H_f^o - T\Delta S_f^o$$

Using standard free energies of formation to calculate the standard free energy of a reaction is analogous to calculating standard enthalpy changes from standard enthalpies of formation using the familiar "products minus reactants" rule:

$$\Delta G^o_{rxn} = \sum m\Delta G^o_f(\text{product}) - \sum n\Delta^o_f(\text{reactants})$$

where m and n are the stoichiometric coefficients of each product and reactant in the balanced chemical equation. A very large negative ΔG° indicates a strong tendency for products to form spontaneously from reactants; it does not, however, necessarily indicate that the reaction will occur rapidly. To make this determination, we need to evaluate the kinetics of the reaction.

The Products Minus Reactants Rule

The ΔG^o of a reaction can be calculated from tabulated ΔG^o_f values using the "products minus reactants" rule.

Example:

Calculate ΔG° for the reaction of isooctane with oxygen gas to give carbon dioxide and water. Use the following data:

- ΔG^o_f(isooctane) = −353.2 kJ/mol,
- $\Delta G^o_f(CO_2)$ = −394.4 kJ/mol, and
- $\Delta G^o_f(H_2O)$ = −237.1 kJ/mol. Is the reaction spontaneous as written?

Given: balanced chemical equation and values of ΔG^o_f for isooctane, CO_2, and H_2O

Asked for: spontaneity of reaction as written.

Strategy:

Use the "products minus reactants" rule to obtain ΔG^o_{rxn}, remembering that ΔG^o_f for an element in its standard state is zero. From the calculated value, determine whether the reaction is spontaneous as written.

Solution:

The balanced chemical equation for the reaction is as follows:

$$C_8H_{18}(1) + \frac{25}{2}O_2(g) \rightarrow 8CO_2(g) + 9H_2O(1)$$

We are given ΔG^o_f values for all the products and reactants except O_2(g). Because oxygen gas is an element in its standard state, $\Delta G^o_f(O_2)$ is zero. Using the "products minus reactants" rule,

$$\Delta G° = \left[8\Delta G^°_f(CO_2) + 9\Delta G^°_f(H_2O)\right] - \left[1\Delta G^°_f(C_8H_{18}) + \frac{25}{2}\Delta G^°_f(O_2)\right]$$
$$= \left[(8\text{ mol})(-394.4\ kJ/\text{mol}) + (9\text{mol})(-237.1\text{kJ}/\text{mol})\right]$$

$$-\left[(1\,\text{mol})(-353.2\,\text{kJ / mol})+\left(\frac{25}{2}\text{mol}\right)(0\,kJ\,/\,mol)\right]$$
$$=-4935.9\ kJ\left(\text{per mol of } C_8H_{18}\right)$$

Because ΔG° is a large negative number, there is a strong tendency for the spontaneous formation of products from reactants (though not necessarily at a rapid rate). Also notice that the magnitude of ΔG° is largely determined by the ΔG°_f of the stable products: water and carbon dioxide.

Calculated values of ΔG° are extremely useful in predicting whether a reaction will occur spontaneously if the reactants and products are mixed under standard conditions. We should note, however, that very few reactions are actually carried out under standard conditions, and calculated values of ΔG° may not tell us whether a given reaction will occur spontaneously under nonstandard conditions. What determines whether a reaction will occur spontaneously is the free-energy change (ΔG) under the actual experimental conditions, which are usually different from ΔG°. If the ΔH and TΔS terms for a reaction have the same sign, for example, then it may be possible to reverse the sign of ΔG by changing the temperature, thereby converting a reaction that is not thermodynamically spontaneous, having $K_{eq} < 1$, to one that is, having a $K_{eq} > 1$, or vice versa. Because ΔH and ΔS usually do not vary greatly with temperature in the absence of a phase change, we can use tabulated values of ΔH° and ΔS° to calculate ΔG° at various temperatures, as long as no phase change occurs over the temperature range being considered.

In the absence of a phase change, neither ΔH nor ΔS vary greatly with temperature.

Example:

Calculate (a) ΔG° and (b) $\Delta G_{300\,°C}$ for the reaction $N_2(g) + 3H_2(g) \rightleftharpoons 2NH_3(g)$, assuming that ΔH and ΔS do not change between 25 °C and 300 °C. Use these data:

- $S°(N_2)$ = 191.6 J/(mol•K),
- $S°(H_2)$ = 130.7 J/(mol•K),
- $S°(NH_3)$ = 192.8 J/(mol•K), and
- $\Delta H^\circ_f(NH_3)$ = −45.9 kJ/mol.

Given: balanced chemical equation, temperatures, S° values, and ΔH°_f for NH_3.

Asked for: ΔG° and ΔG at 300 °C.

Strategy:

- Convert each temperature to kelvins. Then calculate ΔS° for the reaction. Calculate ΔH° for the reaction, recalling that ΔH°_f for any element in its standard state is zero.
- Substitute the appropriate values into equation $\Delta G^\circ = \Delta H^\circ - T\Delta S^\circ$ to obtain ΔG° for the reaction.
- Assuming that ΔH and ΔS are independent of temperature, substitute values into Equation $\Delta G = \Delta H - T\Delta S$ to obtain ΔG for the reaction at 300 °C.

Solution:

1) To calculate ΔG° for the reaction using Equation $\Delta G^{\circ} = \Delta H^{\circ} - T\Delta S^{\circ}$, we must know the temperature as well as the values of ΔS° and ΔH°. At standard conditions, the temperature is 25°C, or 298 K. We can calculate ΔS° for the reaction from the absolute molar entropy values given for the reactants and the products using the "products minus reactants" rule:

$$\text{Ä}S^{\circ}_{rxn} = 2S^{\circ}(NH_3)\text{-}\left[S^{\circ}(N_2)+3S^{\circ}(H_2)\right]$$
$$=\left[2\ \text{mol}\ NH_3 \times 192.8\,\text{J}/(\text{mol}\times\text{K})\right]$$
$$=\left\{\left[1\,\text{mol}\ N_2 \times 191.6\,\text{J}/(\text{mol}\cdot\text{K})\right]+\left[3\,\text{mol}\ H_2 \times 130.7\,\text{J}/(\text{mol}\cdot\text{K})\right]\right\}$$
$$=-198.1\,\text{J}/\text{K}\left(\text{per mole of}\ N_2\right)$$

We can also calculate ΔH° for the reaction using the "products minus reactants" rule. The value of ΔH°_{f} (NH_3) is given, and ΔH°_{f} is zero for both N_2 and H_2:

$$\Delta H^{\circ}_{rxn} = 2\Delta H^{\circ}_{f}(NH_3)-\left[\Delta H^{\circ}_{f}(N_2)+3\Delta H^{\circ}_{f}(H_2)\right]$$
$$=\left[2\times(-45.9\ \text{kJ}/\text{mol})\right]-\left[(1\times 0\,\text{kJ}/\text{mol})+(3\times 0\,\text{kJ}/\text{mol})\right]$$
$$=-91.8\,\text{kJ}\left(\text{per mole of}\ N_2\right)$$

2) Inserting the appropriate values into equation $\Delta G^{\circ} = \Delta H^{\circ} - T\Delta S^{\circ}$:

$$\Delta G^{\circ}_{rxn} = \Delta H^{\circ} - T\Delta S^{\circ} = (-91.8\,\text{kJ})-(298\,\text{K})(-198.1\ \text{J}/\text{K})(1/\text{kJ}/1000\,\text{J}) = -32.7\,\text{kJ}\left(\text{per mol of}\ N_2\right)$$

3) To calculate ΔG for this reaction at 300 °C, we assume that ΔH and ΔS are independent of temperature (i.e., $\Delta H_{300°C}$ = H° and $\Delta S_{300°C}$ = ΔS°) and insert the appropriate temperature (573 K) into Equation $\Delta G = \Delta H - T\Delta S$:

$$\Delta G_{300°C} = \Delta H_{300°C} - (573\,K)(\Delta S_{300°C}) = \Delta H^{\circ} - (573\,K)\Delta S^{\circ}$$
$$(-91.8\ kJ)-(573\,K)(-198.1\ \text{J}/\text{K})(1\text{kJ}/1000\,\text{J}) = 21.7\,\text{kJ}\left(\text{per mole of}\ N_2\right)$$

In this example, changing the temperature has a major effect on the thermodynamic spontaneity of the reaction. Under standard conditions, the reaction of nitrogen and hydrogen gas to produce ammonia is thermodynamically spontaneous, but in practice, it is too slow to be useful industrially. Increasing the temperature in an attempt to make this reaction occur more rapidly also changes the thermodynamics by causing the −TΔS° term to dominate, and the reaction is no longer spontaneous at high temperatures; that is, its K_{eq} is less than one. This is a classic example of the conflict encountered in real systems between thermodynamics and kinetics, which is often unavoidable.

The effect of temperature on the spontaneity of a reaction, which is an important factor in the design of an experiment or an industrial process, depends on the sign and magnitude of both ΔH° and ΔS°. The temperature at which a given reaction is at equilibrium can be calculated by setting ΔG° = 0 in Equation $\Delta G^{\circ} = \Delta H^{\circ} - T\Delta S^{\circ}$.

The reaction of nitrogen and hydrogen gas to produce ammonia is one in which ΔH° and ΔS° are both negative. Such reactions are predicted to be thermodynamically spontaneous at low temperatures but nonspontaneous at high temperatures. Use the data in Example to calculate the temperature at which this reaction changes from spontaneous to nonspontaneous, assuming that ΔH° and ΔS° are independent of temperature.

Given: ΔH° and ΔS°

Asked for: temperature at which reaction changes from spontaneous to nonspontaneous

Strategy: Set ΔG° equal to zero in Equation $\Delta G^{\circ} = \Delta H^{\circ} - T\Delta S^{\circ}$ and solve for T, the temperature at which the reaction becomes nonspontaneous.

Solution:

We calculated that ΔH° is −91.8 kJ/mol of N_2 and ΔS° is −198.1 J/K per mole of N_2, corresponding to ΔG° = −32.7 kJ/mol of N_2 at 25 °C. Thus the reaction is indeed spontaneous at low temperatures, as expected based on the signs of ΔH° and ΔS°. The temperature at which the reaction becomes nonspontaneous is found by setting ΔG° equal to zero and rearranging Equation $\Delta G^{\circ} = \Delta H^{\circ} - T\Delta S^{\circ}$ to solve for T:

$$\Delta G^{\circ} = \Delta H^{\circ} - T\Delta S^{\circ} = 0$$

$$\Delta H^{\circ} = T\Delta S^{\circ}$$

$$T = \frac{\Delta H^{\circ}}{\Delta S^{\circ}} = \frac{(-91.8\,\text{kJ})(1000\,\text{J}/\text{kJ})}{-198.1\,\text{J}/\text{K}} = 463\,\text{K}$$

This is a case in which a chemical engineer is severely limited by thermodynamics. Any attempt to increase the rate of reaction of nitrogen with hydrogen by increasing the temperature will cause reactants to be favored over products above 463 K.

Enthalpy

In thermodynamics, enthalpy (also known as thermodynamic potential) is the sum of the internal energy U and the product of pressure p and volume V of a system,

$$H = U + pV$$

This characteristic function used to be called "heat contents", which is why it is conventionally indicated by *H*. The term "enthalpy" was coined by the Dutch physicist Heike Kamerling Onnes.

The internal energy *U* and the work term *pV* have dimension of energy, in SI units this is joule; the extensive (linear in size) quantity *H* has the same dimension.

Enthalpy has corresponding intensive (size-independent) properties for pure materials. A corresponding intensive property is specific enthalpy, which is enthalpy per mass of substance involved. Specific enthalpy is denoted by a lower case *h*, with dimension of energy per mass (SI unit: joule/kg). If a molecular mass or number of moles involved can be assigned, then another corresponding intensive property is molar enthalpy, which is enthalpy per mole of the compound involved, or alternatively specific enthalpy times molecular mass. There is no universally agreed upon symbol for molar properties, and molar enthalpy has been at times confusingly symbolized by *H*, as in extensive enthalpy. The dimensions of molar enthalpy are energy per number of moles (SI unit: joule/mole).

In terms of intensive properties, specific enthalpy can be correspondingly defined as follows:

$$h = u + pv$$

where,

h = specific enthalpy,

u = specific internal energy,

p = pressure (as before),

v = specific volume = reciprocal of density.

Enthalpy is a function depending on the independent variables that describe the state of the thermodynamic system. Most commonly one considers systems that have three forms of energy contact with their surroundings, namely the reversible and infinitesimal gain of heat, $DQ = TdS$, loss of energy by mechanical work done by the system $-pdV$, and acquiring of substance, $\mu\, dn$. The states of systems with three energy contacts are determined by three independent variables. Although a fairly arbitrary choice of three variables is possible, it is most convenient to consider *H(S,p,n)*, that is, to describe *H* as function of its "natural variables" entropy *S*, pressure *p*, and amount of substance *n*.

In thermodynamics one usually works with differentials. In this case:

$$dH = dU + pdV + Vdp$$

The internal energy dU and the corresponding enthalpy dH are:

$$dU = TdS - pdV + \mu dn \Rightarrow dH = TdS + Vdp + \mu dn$$

The rightmost side is an equation for the characteristic function H in terms of the natural variables S, p, and n.

The first law of thermodynamics can be written—for a system with constant amount of substance—as:

$$DQ = dU + pdV$$

If we keep p constant (an isobaric process) and integrate from state 1 to state 2, we find,

$$\int_1^2 DQ = \int_1^2 dU + \int_1^2 pdV \Rightarrow Q = U_2 - U_1 + p(V_2 - V_1) = (U_2 + pV_2) - (U_1 + pV_1) = H_2 - H_1 = \int_1^2 dH,$$

where symbolically the total amount of heat absorbed by the system, *Q*, is written as an integral.

The other integrals have the usual definition of integrals of functions. The final equation (valid for an isobaric process) is:

$$H_2 - H_1 = Q$$

In other words, if the only work done is a change of volume at constant pressure, $W = p(V_2 - V_1)$, the enthalpy change $H_2 - H_1$ is exactly equal to the heat Q transferred to the system.

As with other thermodynamic energy functions, it is neither convenient nor necessary to determine absolute values of enthalpy. For each substance, the zero-enthalpy state can be some convenient reference state.

Enthalpy Change Accompanying a Change in State

When a liquid vaporizes the liquid must absorb heat from its surroundings to replace the energy taken by the vaporizing molecules in order for the temperature to remain constant. This heat required to vaporize the liquid is called enthalpy of vaporization (or heat of vaporization). For example, the vaporization of one mole of water the enthalpy is given as:

$$\Delta \mathrm{H} = 44.0 \text{ kJ at } 298 \text{ K}$$

When a solid melts, the required energy is similarly called enthalpy of fusion (or heat of fusion). For example, one mole of ice the enthalpy is given as:

$$\Delta \mathrm{H} = 6.01 \text{ kJ at } 273.15 \text{ K}$$

$$\Delta \mathrm{H} = \Delta \mathrm{U} + \mathrm{p}\Delta \mathrm{V}$$

Enthalpy can also be expressed as a molar enthalpy, ΔH_m, by dividing the enthalpy or change in enthalpy by the number of moles. Enthalpy is a state function. This implies that when a system changes from one state to another, the change in enthalpy is independent of the path between two states of a system.

If there is no non-expansion work on the system and the pressure is still constant, then the change in enthalpy will equal the heat consumed or released by the system (q).

$$\Delta H = q$$

This relationship can help to determine whether a reaction is endothermic or exothermic. At constant pressure, an endothermic reaction is when heat is absorbed. This means that the system consumes heat from the surroundings, so q is greater than zero. Therefore according to the second equation, the ΔH will also be greater than zero. On the other hand, an exothermic reaction at constant pressure is when heat is released. This implies that the system gives off heat to the surroundings, so q is less than zero. Furthermore, ΔH will be less than zero.

Effect of Temperature on Enthalpy

When the temperature increases, the amount of molecular interactions also increases. When the number of interactions increase, then the internal energy of the system rises. According to the first equation given, if the internal energy (U) increases then the ΔH increases as temperature rises. We can use the equation for heat capacity and Equation $\Delta H = \Delta U + \Delta PV$ to derive this relationship:

$$C = \frac{q}{\Delta T}$$

Under constant pressure, substitute Equation $\Delta H = q$ into Equation $C = \frac{q}{\Delta T}$:

$$C_p = \left(\frac{\Delta H}{\Delta T}\right)_p$$

Where the subscript P indicates the derivative is done under constant pressure.

The Enthalpy of Phase Transition

Enthalpy can be represented as the standard enthalpy, ΔH°. This is the enthalpy of a substance at standard state. The standard state is defined as the pure substance held constant at 1 bar of pressure. Phase transitions, such as ice to liquid water, require or absorb a particular amount of standard enthalpy:

- Standard Enthalpy of Vaporization $\left(\Delta H^o_{vap}\right)$ is the energy that must be supplied as heat at constant pressure per mole of molecules vaporized (liquid to gas).
- Standard Enthalpy of Fusion $\left(\Delta H^o_{fus}\right)$ is the energy that must be supplied as heat at constant pressure per mole of molecules melted (solid to liquid).
- Standard Enthalpy of Sublimation $\left(\Delta H^o_{sub}\right)$ is the energy that must be supplied as heat at constant pressure per mole of molecules converted to vapor from a solid.

$$\Delta H^o_{sub} = \Delta H^o_{fus} + \Delta H^o_{vap}$$

The enthalpy of condensation is the reverse of the enthalpy of vaporization and the enthalpy of freezing is the reverse of the enthalpy of fusion. The enthalpy change of a reverse phase transition is the negative of the enthalpy change of the forward phase transition. Also the enthalpy change of a complete process is the sum of the enthalpy changes for each of the phase transitions incorporated in the process.

Helmholtz Free Energy

Helmholtz free energy in thermodynamics is a thermodynamic potential which is used to measure the work of a closed system with constant temperature and volume.

The Helmholtz free energy is the observable,

$$F = U - TS.$$

We can get an interpretation of F analogous to that of H as follows. Imagine creating a system in a state with entropy S and energy U in equilibrium with an environment at fixed temperature T. How much energy must be provided? Some of the energy U can be obtained as heat from the environment – this heat is T S, where S is the entropy of the created system. The rest of the energy must be provided as work – this is F. Conversely, if you destroy the system at constant temperature T you need to reduce the entropy to zero via heat transfer T S to the environment, the remaining energy — work which can be extracted (for "free") — is then F. You can think of F as the energy available to do work for a system at temperature T, with energy U, and with entropy S.

A more practical interpretation of F comes when we monitor its change due to an isothermal process. For an isothermal process we have,

$$\Delta F = \Delta U - T\Delta S$$

If the process is also quasi-static, then T ΔS = Q and we have (using the first law):

$$\Delta F = W.$$

Thus, F can be used to measure the work done in a quasi-static isothermal process. F is the observable used to determine work for an isothermal process just as H is used to measure heat for an isobaric process. Note that the work in the above formula is all work of any kind, not just compressional.

If the isothermal process is not quasi-static (e.g., free expansion), then,

$$T\Delta S \geq Q,$$

and we have,

$$\Delta F \leq W.$$

Here W is all work of any kind done on the system. Keep in mind that all these deductions are valid for isothermal processes.

There are a number of important differential relations involving Helmholtz free energy, pressure, entropy, and chemical potential which constitute an alternative form of the thermodynamic identity. Infinitesimally, we have simply from the definition of the Helmholtz free energy:

$$dF = dU - SdT - TdS.$$

Using our original thermodynamic identity built from internal energy, we get a new form of the thermodynamic identity in terms of F:

$$dF = -PdV - SdT + \mu dN.$$

Evidently we have,

$$P = -\left(\frac{\partial F}{\partial V}\right)_{T,N},$$

$$S = -\left(\frac{\partial F}{\partial T}\right)_{V,N},$$

and

$$\mu = \left(\frac{\partial F}{\partial N} \right)_{T,V} .$$

Mathematically, if we express F = F(V, T, N) we can just take partial derivatives to get P, S and μ (expressed as functions of (V, T, N)).

Application of Helmholtz Free Energy

In Equation of State

Pure fluids with high accuracy (like industrial refrigerants) are represented using Helmholtz function as a sum of ideal gas and residual terms.

In Auto-encoder

Auto-encoder is an artificial neural network which is used to encode efficient data. Here Helmholtz energy is used to find the sum of code cost and reconstructed code.

Helmholtz Function

Helmholtz function is a thermodynamic function which is defined as the decrease in the function and is equal to the maximum amount of work which is available during reversible isothermal process.

Difference between Helmholtz Free Energy and Gibbs Free Energy

Helmholtz Free Energy	Gibbs Free Energy
It is defined as the useful work that is obtained from a particular system.	It is defined as the maximum reversible work that is obtained from a particular system.
It is the energy required to create a system at constant temperature and volume.	It is energy required to create a system at constant pressure and temperature.
Helmholtz free energy finds lesser application as the volume of the system should be constant.	Gibbs free energy finds more application as the pressure of the system is constant.

References

- ThermodynamicPotentialsAndThermodynamicParameters: ijemr.net, Retrieved 14 August, 2019
- Energies-and-Potentials/Enthalpy, Supplemental-Modules-(Physical-and-Theoretical-Chemistry): chem.libretexts.org, Retrieved 28 January, 2019
- Helmholtz-free-energy: byjus.com, Retrieved 07 February, 2019

6

Thermodynamic Equations

Mayer's relation, Maxwell's relations, TdS Equations, Bridgman's thermodynamic equations, Clausius Clapeyron equation, Gibbs–Duhem equation, Joule–Thomson effect, etc. are some of the aspects that come within thermodynamic equations. All these aspects of thermodynamic equations have been carefully analyzed in this chapter.

Mayer's Relation

Julius Robert Mayer, a German chemist and physicist, derived a relation between specific heat at constant pressure and the specific heat at constant volume for an ideal gas. He studied the fact that the specific heat capacity of a gas at constant pressure (C_p) is slightly greater than at constant volume (C_v). He reasoned that this C_p is greater than the molar specific heat at constant volume C_v, because energy must now be supplied not only to raise the temperature of the gas but also for the gas to do work because in this case volume changes. According to the Mayer's relation or the Mayer's formula the difference between these two heat capacities is equal to the universal gas constant, thus the molar specific heat at constant pressure is equal:

$$C_p = C_v + R$$

Let us consider one mole of an ideal gas enclosed in a cylinder provided with a frictionless piston of area A. Let P, V and T be the pressure, volume and absolute temperature of gas respectively.

A quantity of heat dQ is supplied to the gas. To keep the volume of the gas constant, a small weight is placed over the piston. The pressure and the temperature of the gas increase to P + dP and T + dT respectively. This heat energy dQ is used to increase the internal energy dU of the gas. But the gas does not do any work (dW = 0).

$$\therefore\ dQ = dU = 1 \times C_v \times dT$$

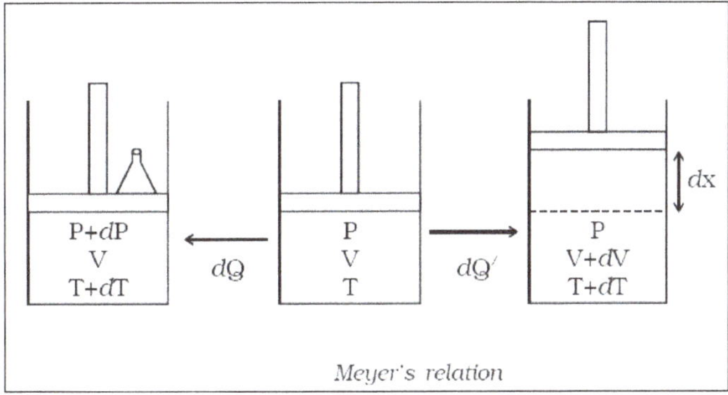

Meyer's relation

The additional weight is now removed from the piston. The piston now moves upwards through a distance dx, such that the pressure of the enclosed gas is equal to the atmospheric pressure P. The temperature of the gas decreases due to the expansion of the gas.

Now a quantity of heat dQ' is supplied to the gas till its temperature becomes T + dT. This heat energy is not only used to increase the internal energy dU of the gas but also to do external work dW in moving the piston upwards.

$$\therefore dQ' = dU + dW$$

Since the expansion takes place at constant pressure,

$$dQ' = CpdT$$

$$\therefore C_p dT = C_v dT + dW$$

Work done, dW = force × distance = P × A × dx,

dW = P dV (since A × dx = dV, change in volume)

$$\therefore C_p dT = C_v dT + P\,dV$$

The equation of state of an ideal gas is:

$$PV = RT$$

Differentiating both the sides:

$$PdV = RdT$$

Substituting equation PdV = RdT in $CpdT = CvdT + P\,dV$,

$$C_p dT = C_v dT + RdT$$

$$C_p = C_v + R$$

This equation is known as Meyer's relation.

Maxwell's Relations

The Maxwell relations are derived from Euler's reciprocity relation. The relations are expressed in partial differential form. The Maxwell relations consists of the characteristic functions: internal energy U, enthalpy H, Helmholtz free energy F, and Gibbs free energy G and thermodynamic parameters: entropy S, pressure P, volume V, and temperature T. Following is the table of Maxwell relations for secondary derivatives:

$$+\left(\frac{\partial T}{\partial V}\right)S = -\left(\frac{\partial P}{\partial S}\right)V = \frac{\partial^2 U}{\partial S \partial V}$$

$$+\left(\frac{\partial T}{\partial V}\right)S=-\left(\frac{\partial P}{\partial S}\right)V=\frac{\partial^2 U}{\partial S\partial V}$$

$$+\left(\frac{\partial T}{\partial P}\right)S=+\left(\frac{\partial V}{\partial S}\right)P=\frac{\partial^2 H}{\partial S\partial P}$$

$$+\left(\frac{\partial S}{\partial V}\right)T=+\left(\frac{\partial P}{\partial T}\right)V=\frac{\partial^2 F}{\partial T\partial V}$$

$$-\left(\frac{\partial S}{\partial P}\right)T=+\left(\frac{\partial V}{\partial T}\right)V=\frac{\partial^2 G}{\partial T\partial P}$$

These are the set of thermodynamics equations which derived from a symmetry of secondary derivatives and from thermodynamic potentials. These relations are named after James Clerk Maxwell, who was a 19th-century physicist.

Derivation of Maxwell's Relations

Maxwell's relations can be derived as:

$$dU=TdS-PdV \text{ (differential form of internal energy)}$$

$$dU=\left(\frac{\partial z}{\partial x}\right)_y dx+\left(\frac{\partial z}{\partial y}\right)_x dy \text{ (total differential form)}$$

$$dz=Mdx+Ndy \text{ (other way of showing the equation)}$$

$$M=\left(\frac{\partial z}{\partial x}\right)_y$$

and

$$N=\left(\frac{\partial z}{\partial y}\right)_x$$

from dU = TdS – PdV

$$T=\left(\frac{\partial U}{\partial S}\right)V$$

and

$$-P=\left(\frac{\partial U}{\partial V}\right)S\frac{\partial}{\partial y}\left(\frac{\partial z}{\partial x}\right)_y=\frac{\partial}{\partial x}\left(\frac{\partial z}{\partial y}\right)_x=\frac{\partial^2 z}{\partial y\partial x}=\frac{\partial^2 z}{\partial x\partial y} \text{ (symmetry of second derivatives)}$$

$$\frac{\partial}{\partial V}\left(\frac{\partial U}{\partial S}\right)V=\frac{\partial}{\partial S}\left(\frac{\partial U}{\partial V}\right)S\left(\frac{\partial T}{\partial V}\right)S=-\left(\frac{\partial P}{\partial S}\right)V$$

Common forms of Maxwell's Relations

Function	Differential	Natural variables	Maxwell Relation
U	dU = TdS – PdV	S, V	$\left(\frac{\partial T}{\partial V}\right)S = -\left(\frac{\partial P}{\partial S}\right)V$
H	dH = TdS + VdP	S, P	$\left(\frac{\partial T}{\partial P}\right)S = -\left(\frac{\partial V}{\partial S}\right)P$
F	dF = -PdV – SdT	V, T	$\left(\frac{\partial P}{\partial T}\right)V = -\left(\frac{\partial S}{\partial V}\right)T$
S	dG = VdP – SdT	P, T	$\left(\frac{\partial V}{\partial T}\right)P = -\left(\frac{\partial S}{\partial P}\right)T$

Where,

T is the temperature

S is the entropy

P is the pressure

V is the volume

U is the internal energy

H is the entropy

G is the Gibbs free energy

F is the Helmholtz free energy

With respect to pressure and particle number, enthalpy and Maxwell's relation can be written as:

$$\left(\frac{\partial \mu}{\partial P}\right)S,N = \left(\frac{\partial V}{\partial N}\right)S,P = \left(\frac{\partial^2 H}{\partial P \partial N}\right)$$

TdS Equations

First TdS Equation

Let entropy S is a function of T and V . i.e.

$$S = S(T,V)$$

$$dS = \left(\frac{\partial S}{\partial T}\right)_V dT + \left(\frac{\partial S}{\partial V}\right)_T dV$$

Multiplying both sides by T, we get:

$$TdS = T\left(\frac{\partial S}{\partial T}\right)_V dT + T\left(\frac{\partial S}{\partial V}\right)_T dV$$

$$TdS = \left(\frac{T\partial S}{\partial T}\right)_V dT + T\left(\frac{\partial S}{\partial V}\right)_T dV$$

$$TdS = \left(\frac{\partial Q}{\partial T}\right)_V dT + T\left(\frac{\partial S}{\partial V}\right)_T dV$$

As specific heat at constant volume $C_V = \left(\frac{\partial Q}{\partial T}\right)_V$:

$$TdS = C_V dT + T\left(\frac{\partial S}{\partial V}\right)_T dV$$

From Maxwell's first equation we get:

$$\left(\frac{\partial S}{\partial V}\right)_T = \left(\frac{\partial P}{\partial T}\right)_V$$

So, we get from $TdS = C_V dT + T\left(\frac{\partial S}{\partial V}\right)_T dV$:

$$TdS = C_V dT + T\left(\frac{\partial P}{\partial T}\right)_V dV$$

This is known as first TdS equation.

Second TdS Equation

Let entropy S is a function of T and P. i.e.

$$S = S(T,P)$$

$$dS = \left(\frac{\partial S}{\partial T}\right)_P dT + \left(\frac{\partial S}{\partial P}\right)_T dP$$

Multiplying both sides by T, we get:

$$TdS = T\left(\frac{\partial S}{\partial T}\right)_P dT + T\left(\frac{\partial S}{\partial P}\right)_T dP$$

$$TdS = \left(\frac{T\partial S}{\partial T}\right)_P dT + T\left(\frac{\partial S}{\partial P}\right)_T dP$$

$$TdS = \left(\frac{\partial Q}{\partial T}\right)_P dT + T\left(\frac{\partial S}{\partial P}\right)_T dP$$

As specific heat at constant pressure $C_P = \left(\frac{\partial Q}{\partial T}\right)_P$:

$$TdS = CpdT + T\left(\frac{\partial S}{\partial P}\right)_T dP$$

From Maxwell's second equation we get:

$$\left(\frac{\partial S}{\partial P}\right)_T = -\left(\frac{\partial V}{\partial T}\right)_P$$

So, we get from:

$$TdS = C_V dT + T\left(\frac{\partial S}{\partial V}\right)_T dV$$

$$TdS = CpdT - T\left(\frac{\partial V}{\partial T}\right)_P dP$$

This is known as Second TdS equation.

Third TdS Equation

Let entropy S is a function of V and P. i.e.

$$S = S(V,P)$$

$$dS = \left(\frac{\partial V}{\partial V}\right)_P dT + \left(\frac{\partial S}{\partial P}\right)_V dP$$

Multiplying both sides by T, we get:

$$TdS = T\left(\frac{\partial S}{\partial V}\right)_P dV + T\left(\frac{\partial S}{\partial P}\right)_V dP$$

$$TdS = \left(\frac{T\partial S}{\partial V}\right)_P dV + T\left(\frac{\partial S}{\partial P}\right)_V dP$$

$$TdS = \left(\frac{\partial Q}{\partial V}\right)_P dV + T\left(\frac{\partial S}{\partial T}\right)_V dP$$

$$TdS = \left(\frac{\partial Q}{\partial T}\right)_P \left(\frac{\partial T}{\partial V}\right)_P dV + T\left(\frac{\partial S}{\partial T}\right)_V \left(\frac{\partial T}{\partial P}\right)_V dP$$

$$TdS = C_P\left(\frac{\partial T}{\partial V}\right)_P dV + C_V\left(\frac{\partial T}{\partial P}\right)_V dP$$

This is known as third TdS equation.

Bridgman's Thermodynamic Equations

In thermodynamics, Bridgman formulas or Bridgman equations are twenty-eight partial derivatives of the ten essential thermodynamic quantities, involving first and second derivatives, expressed in terms of measureable variables. These formulas were systematically derived by American physicist Percy Bridgman in 1914 using what has come to be known as the Bridgman formula method.

Bridgman, in short, presented a clean and organized way of deriving thermodynamic expressions from the partial derivatives of the ten (or ten) main thermodynamic quantities (or variables), P, V, T, E, H, S, A, and F, of which there are 8 × 7 × 6 first (partial) derivatives.

In his 1925 "full version", in contrast to the 1914 "shorter version", of his systematic collection of thermodynamic formulas, he presented the first derivatives of the 10 fundamental quantities, which amounts to 720 equations, organized into 10 groups, based on which variable was held constant during the differentiation.

Bridgman devised a system that allows for the derivation of any of these first partial derivatives in terms of three quantities which are, in general, capable of experimental determination, namely (∂V/∂T)P, (∂V/∂P)T, and (∂H/∂T)P, i.e. CP. The twenty-eight essential formulas derived using this methodology are:

$$(\partial T)_P = -(\partial P)_r = 1$$

$$(\partial V)_P = -(\partial P)_V = (\partial V / \partial T)_P$$

$$(\partial S)_P = -(\partial P)_S = C_P / T$$

$$(\partial E)_P = -(\partial P)_E = C_P - P(\partial V / \partial T)_P$$

$$(\partial H)_P = -(\partial P)_H = C_P$$

$$(\partial F)_P = -(\partial P)_F = -S$$

$$(\partial A)_P = -(\partial P)_A = -S - P(\partial V / \partial T)_P$$

$$(\partial V)_T = -(\partial T)_V = -(\partial V / \partial P)_T$$

$$(\partial S)_T = -(\partial T)_S = (\partial V / \partial T)_P$$

$$(\partial E)_T = -(\partial T)_E = T(\partial V / \partial T)_P + P(\partial V / \partial P)_T$$

$$(\partial H)_T = -(\partial T)_H = -V + T(\partial V / \partial T)_r$$

$$(\partial F)_T = -(\partial T)_F = -V$$

$$(\partial A)_T = -(\partial T)_A = P(\partial V / \partial P)_T$$

$$(\partial S)_V = -(\partial V)_S = C_P(\partial V / \partial P)_T + T(\partial V / \partial T)_r^2$$

$$(\partial E)_V = -(\partial V)_E = C_P(\partial V / \partial P)_T + T(\partial V / \partial T)_P^2$$

$$(\partial H)_V = -(\partial V)_H = C_P(\partial V / \partial P)_T + T(\partial V / \partial T)_r^2 - V(\partial V / \partial T)_P$$

$$(\partial F)_V = -(\partial V)_F = -V(\partial V/\partial T)_P - S(\partial V/\partial P)_T$$
$$(\partial A)_V = -(\partial V)_A = -S(\partial V/\partial P)_T$$
$$(\partial E)_S = -(\partial S)_E = PC_P(\partial V/\partial T)_T/T + P(\partial V/\partial T)_P^2$$
$$(\partial H)_S = -(\partial S)_H = -VC_P/T$$
$$(\partial F)_S = -(\partial S)_F = -VC_P/T + S(\partial V/\partial T)_r$$
$$(\partial A)_S = -(\partial S)_A = PC_P(\partial V/\partial P)_T/T + P(\partial V/\partial T)_P^2 + S(\partial V/\partial T)_P$$
$$(\partial H)_E = -(\partial E)_H = -V\left[C_P - P(\partial V/\partial T)_P\right] - P\left[C_P(\partial V/\partial P)_T + T(\partial V/\partial T)_P^2\right]$$
$$(\partial F)_E = -(\partial E)_F = -V\left[C_P - P(\partial V/\partial T)_P\right] + S\left[T(\partial V/\partial T)_P + P(\partial V/\partial P)_T\right]$$
$$(\partial A)_E = -(\partial E)A = P\left[C_P(\partial V/\partial P)_T + T(\partial V/\partial T)_P^2\right]$$
$$(\partial F)_H = -(\partial H)_F = -V(C_P + S) + TS(\partial V/\partial T)_P$$
$$(\partial A)_H = -(\partial H)_A = -\left[S + P(\partial V/\partial T)_P\right]\left[V - T(\partial V/\partial T)_P\right] + P(\partial V/\partial P)_T$$
$$(\partial A)_F = -(\partial F)_A = -S\left[V + P(\partial V/\partial P)_T\right] - PV(\partial V/\partial T)_P.$$

Ostwald–Freundlich Equation

The Ostwald–Freundlich equation governs boundaries between two phases; specifically, it relates the surface tension of the boundary to its curvature, the ambient temperature, and the vapor pressure or chemical potential in the two phases.

The Ostwald–Freundlich equation for a droplet or particle with radius R is:

$$\frac{p}{p_{\text{eq}}} = \exp\left(\frac{R_{\text{critical}}}{R}\right)$$

$$R_{critical} = \frac{2 \cdot \gamma \cdot V_{\text{atom}}}{k_{\text{B}} \cdot T}$$

V_{atom} : Atomic volume

k_{B} : Boltzmann constant

γ : Surface tension ($J \cdot m^{-2}$)

p_{eq} : Equilibrium partial pressure (or chemical potential or concentration)

p : Partial pressure (or chemical potential or concentration)

T : Absolute temperature

One consequence of this relation is that small liquid droplets (i.e., particles with a high surface curvature) exhibit a higher effective vapor pressure, since the surface is larger in comparison to the volume.

Another notable example of this relation is Ostwald ripening, in which surface tension causes small precipitates to dissolve and larger ones to grow. Ostwald ripening is thought to occur in the formation of orthoclase megacrysts in granites as a consequence of subsolidus growth.

In 1871, Lord Kelvin (William Thomson) obtained the following relation governing a liquid-vapor interface:

$$p(r_1, r_2) = P - \frac{\gamma \rho_{\text{vapor}}}{(\rho_{\text{liquid}} - \rho_{\text{vapor}})} \left(\frac{1}{r_1} + \frac{1}{r_2} \right)$$

where:

$p(r)$: vapor pressure at a curved interface of radius r,

P: vapor pressure at flat interface $(r = \infty) = p_{eq}$,

γ: surface tension,

ρ_{vapor}: density of vapour,

ρ_{liquid}: density of liquid,

r_1, r_2 :radii of curvature along the principal sections of the curved interface.

In his dissertation of 1885, Robert von Helmholtz (son of the German physicist Hermann von Helmholtz) derived the Ostwald–Freundlich equation and showed that Kelvin's equation could be transformed into the Ostwald–Freundlich equation. The German physical chemist Wilhelm Ostwald derived the equation apparently independently in 1900; however, his derivation contained a minor error which the German chemist Herbert Freundlich corrected in 1909.

Derivation from Kelvin's Equation

According to Lord Kelvin's equation of 1871,

$$p(r_1, r_2) = P - \frac{\gamma \rho_{\text{vapor}}}{(\rho_{\text{liquid}} - \rho_{\text{vapor}})} \left(\frac{1}{r_1} + \frac{1}{r_2} \right).$$

If the particle is assumed to be spherical, then $r = r_1 = r_2$; hence,

$$p(r) = P - \frac{2\gamma \rho_{\text{vapor}}}{(\rho_{\text{liquid}} - \rho_{\text{vapor}})r}.$$

Kelvin defined the surface tension γ as the work that was performed per unit area by the interface rather than on the interface; hence his term containing γ has a minus sign. In what follows, the surface tension will be defined so that the term containing γ has a plus sign.

since $\rho_{\text{liquid}} \gg \rho_{\text{vapor}}$,

then $\rho_{\text{liquid}} - \rho_{\text{vapor}} \approx \rho_{\text{liquid}}$; hence,

$$p(r) \approx P + \frac{2\gamma \rho_{\text{vapor}}}{\rho_{\text{liquid}} \cdot r}$$

Assuming that the vapor obeys the ideal gas law, then,

$$\rho_{\text{vapor}} = \frac{m_{\text{vapor}}}{V} = \frac{MW \cdot n}{V} = \frac{MW \cdot P}{RT} = \frac{MW \cdot P}{N_{\text{A}} k_{\text{B}} T}$$

where:

m_{vapor} : mass of a volume V of vapour,

MW : molecular weight of vapour,

n : number of moles of vapor in volume V of vapour,

N_{A} : Avogadro constant,

R : ideal gas constant = $N_{\text{A}} k_{\text{B}}$.

Since $\frac{MW}{N_{\text{A}}}$: mass of one molecule of vapor or liquid, then:

$$\frac{\left(\frac{MW}{N_{\text{A}}}\right)}{\rho_{\text{liquid}}} = \text{volume of one molecule} = V_{\text{molecule}}.$$

Hence,

$$p(r) \approx P + \frac{2\gamma V_{\text{molecule}} P}{k_{\text{B}} T r} = P + \frac{R_{\text{critical}} P}{r},$$

where $R_{\text{critical}} = \frac{2\gamma V_{\text{molecule}}}{k_{\text{B}} T}$. Thus,

$$\frac{p(r) - P}{P} \approx \frac{R_{\text{critical}}}{r}.$$

Since,

$$\frac{p(r)}{P} = 1 - \frac{P - p(r)}{P},$$

then,

$$\log\left(\frac{p(r)}{P}\right)=\log\left(1-\frac{P-p(r)}{P}\right).$$

Since $p(r)\approx P$, Then $\frac{P-p(r)}{P}\ll 1$.

If $x\ll 1$, then:

$$\log(1-x)\approx -x.$$

Hence,

$$\log\left(\frac{p(r)}{P}\right)\approx\frac{p(r)-P}{P}.$$

Therefore,

$$\log\left(\frac{p(r)}{P}\right)\approx\frac{R_{\text{critical}}}{r},$$

which is the Ostwald–Freundlich equation.

Generalization of the Gibbs–Kelvin–Köhler and Ostwald–Freundlich Equations

A derivation of chemical equilibrium equations for a spherical thin film of solution around a soluble solid nanoparticle is presented. The equations obtained generalize the Gibbs–Kelvin–Köhler and Ostwald–Freundlich equations for a soluble particle immersed in the bulk phase. The generalized equations describe the dependence of the chemical potentials of a condensate and dissolved nanoparticle matter in the thin solution film, the condensate saturated pressure, and the solubility of the nanoparticle matter on the film thickness, and the nanoparticle size with account of the disjoining pressure of the liquid film.

Gibbs' relationship:

$$\mu_1=\mu_{1,\infty}+\frac{2\sigma^{\alpha\beta}v_1^{\alpha}}{R}$$

Is basic for the thermodynamics of homogeneous and heterogeneous nucleations. Equation $\mu_1=\mu_{1,\infty}+\frac{2\sigma^{\alpha\beta}v_1^{\alpha}}{R}$ determines how the condensate chemical potential μ_1 in a spherical embryo of phase α, surrounded by metastable phase β, depends on radius R of the embryo surface at a fixed temperature T. Here $\mu_{1,\infty}$ is the condensate chemical potential at the equilibrium of phases α and β with a flat interface, v_1^{α} is the volume per condensate molecule in phase α under the assumption that the phase is incompressible, and $\sigma^{\alpha\beta}$ is the surface tension related to the embryo surface of tension; all the quantities are taken at the same temperature T. If phase β is an ideal gas and phase α is a liquid (correspondingly, the embryo is a droplet), equation, $\mu_1 = \mu_{1,\infty} + \frac{2\sigma^{\alpha\beta} v_1^\alpha}{R}$ can be rewritten in the form of the Gibbs–Kelvin equation,

$$k_B T \, \mathrm{In} \frac{p_{1,R}^\beta}{p_{1,\infty}^\beta} = \frac{2\sigma^{\alpha\beta} v_1^\alpha}{R}$$

For the equilibrium partial pressure $p_{1,R}^\beta$ of the condensate vapor ($p_{1,\infty}^\beta$ is a value of $p_{1,R}^\beta$ for a flat interface between phases α and β and k_B is the Boltzmann constant). If the droplet includes, in addition to the condensate (component 1) playing the role of a solvent, also a solute (component 2) with a relative bulk concentration $x \equiv c_2^\alpha / c_2^\alpha$ [c_i is the number of molecules of component i (i =1, 2) per unit volume; the variable x coincides with the molar fraction for dilute solutions], then the Gibbs–Kelvin equation transforms into the Gibbs–Kelvin–Köhler equation,

$$k_B \, In \frac{p_{1,R}^\beta}{p_{1,\infty}^\beta} = \frac{2\sigma^{\alpha\beta} v_1^\alpha}{R} - k_B T x.$$

A relationship for the solubility x_R of a small solid particle (phase γ) of radius R in the ideal bulk solution (phase α) is analogous to the Gibbs–Kelvin equation and is called the Ostwald–Freundlich equation,

$$k_B T \;\; \mathrm{In} = \frac{x_R}{x_\infty} = \frac{2\sigma^{\alpha\gamma} v_2^\gamma}{R}.$$

Here x_∞ is the relative concentration of solute in phase α at the equilibrium of phases α and γ with a flat interface, v_2^γ is the volume per molecule in the solid particle, and $\sigma^{\alpha\gamma}$ is the surface tension of the solid particle at its boundary with phase α ; all the quantities being taken at the same temperature T.

The problem of description of chemical equilibrium of a droplet condensing out the vapor-gas environment around partially dissolving solid nanoparticle (the so-called deliquescence problem-recently roused a considerable interest in view of new experimental technique for studying ultrafine aerosols. Such a droplet consists of a spherical liquid film of saturated solution of the particle matter and the solid core) the remainder of the particle. It was experimentally proven that, in the case of high solubility of the solid matter, the droplet with a soluble solid core can be stable in an undersaturated vapor. It is theoretically expected8 that, in the case of low solubility, the droplet can be stable also in slightly supersaturated vapor. As a rule, the liquid film in a stable droplet is thin, with thickness much smaller than the radius of a solid core.

As was first shown theoretically, the existence of thin stable films on partially dissolved cores can be explained using the concept of the film disjoining pressure. The disjoining pressure takes into account the difference of the normal component of the pressure tensor in a thin liquid film from the pressure in the bulk liquid phase of the same nature and at same values of temperature and chemical potentials of components as in the film. The disjoining pressure is produced by overlapping of the surface layers related to the boundaries with the core and the vapor-gas medium. This overlapping is a result of long-ranged electrostatic and molecular forces (the electrostatic and molecular contributions to the disjoining pressure and near-to-wall ordering)the structural contribution to the disjoining pressure.

Earlier, the applications and extensions of the Gibbs– Kelvin–Köhler equation for the condensate chemical potential in the film on soluble solid core and the Ostwald– Freundlich equation for the solubility of solid matter in the film with account of disjoining pressure were considered. The relationships found referred to the quasiflat films of a dilute solution around partially soluble macroscopic condensation nuclei, with the surface tensions at the film boundaries taken the same as for a film of pure condensate. The Ostwald–Freundlich equation was taken in the form expressed in equation,

$$k_B T \text{ In} = \frac{x_R}{x_\infty} = \frac{2\sigma^{\alpha\gamma} v_2^\gamma}{R}.,$$

and the difference between the partial molecular volumes of condensate and solute was neglected. An account of this difference and an extension of the Ostwald–Freundlich equation for a film of ideal solution on a nanoscaled condensation nucleus were presented in more recent papers.

Expanding recently the definition of the disjoining pressure to the case of thin liquid films on small solid particles, we showed the condition of mechanical equilibrium in a spherical film with overlapping surface layers to be written as:

$$p^\alpha - p^\beta = \frac{2\sigma^{\alpha\beta}}{R} - \left(p_N - p^\alpha\right)\frac{R_n^2}{R^2},$$

Where p^α and p^β are the total pressures in the mother bulk phase of the film and the vapor phase at equilibrium with the film, R is the outer film radius, p_N is the normal component of the pressure tensor in the film at the boundary with the particle, and R_n is the particle radius. This refinement is important in the case of nanoscaled condensation nuclei, but it was earlier not recognized in the analysis of Gibbs–Kelvin–Köhler and Ostwald–Freundlich equations for thin films. Besides, when considering the conditions of equilibrium of a solid particle, one should take into account the difference between the thermodynamically and mechanically defined surface tensions of the particle and the fact that the chemical potential of the particle matter cannot be the same in solid particle and solution even at true equilibrium. In addition, the surface tensions at the film boundaries depend, for thin films, on the film thickness and are related to the disjoining pressure. Finally, because a thin film is inhomogeneous and the bulk density cannot be achieved anywhere within the film, the question arise what one should understand as a solute concentration entering Gibbs–Kelvin–Köhler and Ostwald–Freundlich equations for a thin film.

The problems noted indicate that some corrections to the derivation of Gibbs–Kelvin–Köhler and Ostwald–Freundlich equations are required in the case of thin spherical solution films on soluble solid nanoparticles, and we elaborate such corrections in this work. Resulting generalized equations will allow us to find the dependence on the film thickness and the nanoparticle size for the chemical potentials of all species, the saturated pressure of condensate vapor, and the solubility of the nanoparticle.

Relationships for the Chemical Potentials of Condensate and Dissolved Core Matter in a Curved Thin Film

Let us consider a droplet in the form of a liquid film around a spherical nonvolatile one-component solid core, the drop being in equilibrium with a surrounding vapor-gas medium. The solid matter is soluble in the film, so that the evaporation-condensation process between the drop and its surroundings can be accompanied by the dissolution-deposition process between the solid core and the film. We mark quantities referred to liquid, gas, and solid phases with superscripts α, β, and γ, respectively. It should be noted at once that the state of a bulk phase can be unattainable inside the core and the enveloping film in view of their small dimensions. It does not matter for the core because only surface monolayer plays the determinative role for solids, although we may use also quantities referred to the solid reference bulk phase. The properties of liquid bulk phase α can differ from those for the thin film. However, phase α is a mother phase for the film and, when taken at the same values of temperature and chemical potentials, is of fundamental importance for description of the film.

We mark the condensing component by subscript 1 and the dissolving component by subscript 2. Let the outer radius of a droplet with a core be R and the core radius R_n. Our first task will be finding an isothermal dependence of the condensate and solute chemical potentials in the film on radii R and R_n.

Let us start with the condensate chemical potential μ_1^α in the phase α being a uniform solution of the core matter in the condensate with relative concentration x at pressure p^α and temperature T. As is known, the dissolution of foreign matter always decreases the chemical potential of a solvent, the effect being the stronger, the higher the solute concentration. This effect can be described in terms of the osmotic pressure $\pi(x)$ under the assumption of incompressibility of phase α as,

$$\mu_1^\alpha\left(p^\alpha,x\right)=\mu_{1,0}^\alpha\left(p^\alpha\right)-v_1^\alpha\pi(x),$$

where $\mu_{1,0}^\alpha$ is the molecular chemical potential in a pure condensate at the same pressure and temperature, and v_1^α is the partial molecular volume of the condensate in the solution.

Let $\Delta p\equiv p^\alpha-p^\beta$ be the total pressure difference for the liquid and gas phases. Then equation, $\mu_1^\alpha\left(p^\alpha,x\right)=\mu_{1,0}^\alpha\left(p^\alpha\right)-v_1^\alpha\pi(x)$, with the same accuracy can be rewritten in the form:

$$\mu_1^\alpha\left(p^\beta+\Delta p,x\right)=\mu_{1,0}^\alpha\left(p^\beta\right)+v_1^\alpha\left[\Delta p-\pi(x)\right].$$

If the solution film around the core is thin, then overlapping the surface layers produces the disjoining pressure Π in the film. We define Π as,

$$\Pi\equiv p_N-p^\alpha,$$

where p^α is pressure in the mother phase of the film at the same temperature and chemical potentials as in the film and p_N is the normal component of the pressure tensor in the film at its inner boundary with the core. Thus, when phase α is at equilibrium with the spherical thin film, the chemical potentials of condensate and solvent in the phase α and the film coincide, but in view of equation, $p^\alpha - p^\beta = \frac{2\sigma^{\alpha\beta}}{R} - \left(p_N - p^\alpha\right)\frac{R_n^2}{R^2}$, and $\Pi \equiv p_N - p^\alpha$, pressure p^α in phase α differs from the external pressure p^β by the quantity,

$$\Delta p \equiv p^\alpha - p^\beta = \frac{2\sigma^{\alpha\beta}}{R} - \frac{R_n^2}{R^2}\Pi,$$

where $\sigma^{\alpha\beta}$ is the surface tension at the boundary between the film and the gas phase. Disjoining pressure Π equals zero for a thick film, and equation, $\Delta p \equiv p^\alpha - p^\beta = \frac{2\sigma^{\alpha\beta}}{R} - \frac{R_n^2}{R^2}\Pi$, reduces to the Laplace equation in this case. In the general case, pressure Π is evident to depend on concentration x of the solute in the bulk solution kept in equilibrium with the film. In its turn, concentration x is determined by the values of chemical potentials in phase α.

Taking into account the condition of equality of the condensate chemical potentials in phase α and the film, we find, using equation,

$$\mu_1^\alpha\left(p^\beta + \Delta p, x\right) = \mu_{1,0}^\alpha\left(p^\beta\right) + v_1^\alpha\left[\Delta p - \pi(x)\right].$$

and $\Delta p \equiv p^\alpha - p^\beta = \frac{2\sigma^{\alpha\beta}}{R} - \frac{R_n^2}{R^2}\Pi$, , the following relationship for the condensate chemical potential μ_1 in the film:

$$\begin{aligned}\mu_1 &= \mu_1^\alpha\left(p^\beta + \Delta p, x\right)\\ &= \mu_{1,0}^\alpha\left(p^\beta\right) + \frac{2\sigma^{\alpha\beta}v_1^\alpha}{R} - v_1^\alpha\left[\frac{R_n^2}{R^2}\Pi + \pi(x)\right].\end{aligned}$$

In view of the approximate equality $\mu_{1,0}^\alpha\left(p^\beta\right) \simeq \mu_{1,\infty}$ [the difference $p^\beta - p_{1,\infty}^\beta$ is assumed to be much smaller than Δp or $\pi(x)$ for nanosized droplets], equation,

$$\mu_1 = \mu_1^\alpha\left(p^\beta + \Delta p, x\right) = \mu_{1,0}^\alpha\left(p^\beta\right) + \frac{2\sigma^{\alpha\beta}v_1^\alpha}{R} - v_1^\alpha\left[\frac{R_n^2}{R^2}\Pi + \pi(x)\right].$$

generalizes the Gibbs equation $\mu_1 = \mu_{1,\infty} + \frac{2\sigma^{\alpha\beta}v_1^\alpha}{R}$ to the case of thin spherical solution film on a solid core.

It is not hard to obtain, in a similar way, an expression for the solute chemical potential μ_2. Using the condition of equality of the solute chemical potentials in phase α and the film, we find, with the help of equation, $\Delta p \equiv p^{\alpha} - p^{\beta} = \frac{2\sigma^{\alpha\beta}}{R} - \frac{R_n^2}{R^2}\Pi$, the following relationship:

$$\mu_2 = \mu_2^{\alpha}\left(p^{\beta} + \Delta p, x\right) = \mu_2^{\alpha}\left(p^{\beta}, x\right) + \frac{2\sigma^{\alpha\beta} v_2^{\alpha}}{R} - v_2^{\alpha}\frac{R_n^2}{R^2}\Pi,$$

Where μ_2^{α} is the solute chemical potential and v_2^{α} is the solute partial molecular volume in phase α. On the other hand, the chemical potential should be equal to the chemical potential of the core matter at the core surface. As is known, the equality of the chemical potentials of the core matter inside the spherical solid particle and in the solution breaks. Nevertheless, if we know the value of the solute chemical potential in the solution, we know the value of the chemical potential of the solid at the core surface. Using equation, $k_B T\ In\frac{a(x)}{a_{\infty}} = v_2^{\gamma}\left(\Pi + \frac{2\sigma^{\alpha\gamma}}{R_n} + \left.\frac{\partial\sigma^{\alpha\gamma}}{\partial R_n}\right|_{h=R-R_n}\right) + \left(v_2^{\gamma} - v_2^{\alpha}\right)\left(\frac{2\sigma^{\alpha\beta}}{R} - \frac{R_n^2}{R^2}\Pi\right)$.

and assuming the core incompressible, one may write the following expression for the chemical potential $\mu_2^{\gamma}\left(p^{\beta}\right)$ of the core matter at the surface of the core under external pressure p^{β}:

$$\mu_2^{\gamma}\left(p^{\beta}\right) = \mu_{2,\infty}^{\gamma} + v_2^{\gamma}\frac{2\sigma_{R_n}^{\alpha\gamma}}{R_n} + v_2^{\gamma}\frac{d\sigma_{R_n}^{\alpha\gamma}}{dR_n},$$

Where $\mu_{2,\infty}^{\gamma}$ is the chemical potential of the solid matter at the flat interface between the solid substrate and the bulk solution with pressure p^{β} and concentration x_{∞} and $\sigma_{R_n}^{\alpha\gamma}$ is the thermodynamic surface tension at the core-solution boundary. Note that chemical potential $\mu_{2,\infty}^{\gamma}$ can be represented as

$$\mu_{2,\infty}^{\gamma} = v_2^{\gamma}\left(f + P_{N,\infty}\right) = \mu_{2,\infty}^{\alpha},$$

Where f is the free energy density in the solid phase,

$$P_{N,\infty} = P^{\beta}$$

Is the normal component of the pressure tensor in the film at the boundary with the flat substrate, and

$$\mu_{2,\infty}^{\alpha} \equiv \mu_2^{\alpha}\left(p^{\beta}, x_{\infty}\right).$$

Changing the normal component of the pressure tensor at the substrate-solution boundary by quantity $\Delta p_N \equiv p_N - p^{\beta}$,, where p_N corresponds, as before, to the value at the inner boundary of the spherical film around the solid core, we obtain from equation, $\mu_2^{\gamma}\left(p^{\beta}\right) = \mu_{2,\infty}^{\gamma} + v_2^{\gamma}\frac{2\sigma_{R_n}^{\alpha\gamma}}{R_n} + v_2^{\gamma}\frac{d\sigma_{R_n}^{\alpha\gamma}}{dR_n}$, $\mu_{2,\infty}^{\gamma} = v_2^{\gamma}\left(f + P_{N,\infty}\right) = \mu_{2,\infty}^{\alpha}$, $P_{N,\infty} = P^{\beta}$, $\mu_{2,\infty}^{\alpha} \equiv \mu_2^{\alpha}\left(p^{\beta}, x_{\infty}\right)$. the following useful relationship:

$$\mu_2 = \mu_2^{\gamma}\left(p^{\beta} + \Delta p_N\right)$$

$$= \mu_2^{\alpha}\left(p^{\beta}, x_{\infty}\right) + v_2^{\gamma}\left(\Delta p_N + \frac{2\sigma^{\alpha\gamma}}{R_n} + \left.\frac{\partial\sigma^{\alpha\gamma}}{\partial R_n}\right|_{h=R-R_n}\right),$$

Which is hard to get for free by another approach. In this procedure surface tension $\sigma_{R_n}^{\alpha\gamma}$ turns in equation,

$$\mu_2 = \mu_2^{\gamma}\left(p^{\beta} + \Delta p_N\right) = \mu_2^{\alpha}\left(p^{\beta}, x_{\infty}\right) + v_2^{\gamma}\left(\Delta p_N + \frac{2\sigma^{\alpha\gamma}}{R_n} + \left.\frac{\partial \sigma^{\alpha\gamma}}{\partial R_n}\right|_{h=R-R_n}\right),$$

into surface tension $\sigma^{\alpha\gamma}$ at the inner boundary of the film, the derivative $d\sigma_{R_n}^{\alpha\gamma} / dR_n$ at a fixed state of the bulk solution into derivative $\left(\partial\sigma^{\alpha y} / \partial R_n\right)\Big|_{h=R-R_n}$ at a fixed film thickness $h = R - R_N$. Therewith, the quantity $\Delta p_N \equiv p_N - p^{\beta}$ has a sense of the difference of normal components of the pressure tensor at the inner and outer film boundaries. Note that the state of a curved thin film depends not only on the film thickness but also on the curvatures of each of the film boundaries as well because the curvatures can be varied independently. The fact that we fix only the film thickness means that the solid core is assumed to be large in comparison with the molecular size. Thus, the condition of a fixed film thickness which is strictly valid for a flat film serves here as an approximation.

Using the equality:

$$\Delta p_N \equiv p_N - p^{\beta} = \Pi + \frac{2\sigma^{\alpha\beta}}{R} - \frac{R_n^2}{R^2}\Pi,$$

which follows from equation $\Pi \equiv p_N - p^{\alpha}$, and $\Delta p \equiv p^{\alpha} - p^{\beta} = \frac{2\sigma^{\alpha\beta}}{R} - \frac{R_n^2}{R^2}\Pi$, we find from equation $\mu_2 = \mu_2^{\gamma}\left(p^{\beta} + \Delta p_N\right) = \mu_2^{\alpha}\left(p^{\beta}, x_{\infty}\right) + v_2^{\gamma}\left(\Delta p_N + \frac{2\sigma^{\alpha\gamma}}{R_n} + \left.\frac{\partial \sigma^{\alpha\gamma}}{\partial R_n}\right|_{h=R-R_n}\right)$,

$$\mu_2 = \mu_2^{\alpha}\left(p^{\beta}, x_{\infty}\right) + v_2^{\gamma}\left(\Pi + \frac{2\sigma^{\alpha\beta}}{R} - \frac{R_n^2}{R^2}\Pi + \frac{2\sigma^{\alpha\gamma}}{R_n} + \left.\frac{\partial \sigma^{\alpha\gamma}}{\partial R_n}\right|_{h=R-R_n}\right).$$

This expression for the solute chemical potential allows us to derive an equation for the core activity and solubility.

The Core Solubility and the Saturated Vapor Pressure of the Condensate as Functions of the Film and Core Radii

The standard expression for the chemical potential $\mu_2^{\alpha}\left(p^{\beta}, x\right)$ in phase α, which is the bulk solution of the solid core matter in the condensate with relative concentration x and activity $a(x)$ at pressure p^{β} and temperature T, has a form,

$$\mu_2^{\alpha}\left(p^{\beta},x\right)=\mu_2^{*}\left(p^{\beta}\right)+k_B T\ In\ a\left(x\right),$$

where μ_2^{*} is the standard part of the solute chemical potential (as $x \to 0$). The analogous expression for $\mu_{2,\infty}^{\alpha}$ can be written as,

$$\mu_{2,\infty}^{\alpha}=\mu_2^{*}\left(p^{\beta}\right)+k_B T\,\mathrm{In}\,\alpha_{\infty},$$

With $a_{\infty}\equiv a\left(x_{\infty}\right)$. Substituting equation $\mu_2^{\alpha}\left(p^{\beta},x\right)=\mu_2^{*}\left(p^{\beta}\right)+k_B T\ In\ a\left(x\right)$, into equation,

$$\mu_2=\mu_2^{\alpha}\left(p^{\beta}+\Delta p,x\right)=\mu_2^{\alpha}\left(p^{\beta},x\right)+\frac{2\sigma^{\alpha\beta}v_2^{\alpha}}{R}-v_2^{\alpha}\frac{R_n^2}{R^2}\Pi,$$

and correspondingly, equation $\mu_{2,\infty}^{\alpha}=\mu_2^{*}\left(p^{\beta}\right)+k_B T\,\mathrm{In}\,\alpha_{\infty}$, into equation,

$$\mu_2=\mu_2^{\alpha}\left(p^{\beta},x_{\infty}\right)+v_2^{\gamma}\left(\Pi+\frac{2\sigma^{\alpha\beta}}{R}-\frac{R_n^2}{R^2}\Pi+\frac{2\sigma^{\alpha\gamma}}{R_n}+\left.\frac{\partial\sigma^{\alpha\gamma}}{\partial R_n}\right|_{h=R-R_n}\right)$$

and equating the righthand sides of equation,

$$\mu_2=\mu_2^{\alpha}\left(p^{\beta}+\Delta p,x\right)=\mu_2^{\alpha}\left(p^{\beta},x\right)+\frac{2\sigma^{\alpha\beta}v_2^{\alpha}}{R}-v_2^{\alpha}\frac{R_n^2}{R^2}\Pi,$$

and

$$\mu_2=\mu_2^{\alpha}\left(p^{\beta},x_{\infty}\right)+v_2^{\gamma}\left(\Pi+\frac{2\sigma^{\alpha\beta}}{R}-\frac{R_n^2}{R^2}\Pi+\frac{2\sigma^{\alpha\gamma}}{R_n}+\left.\frac{\partial\sigma^{\alpha\gamma}}{\partial R_n}\right|_{h=R-R_n}\right)$$

we obtain an equation for the solute activity as a function of radii R_N and R,

$$k_B T\ In\frac{a(x)}{a_{\infty}}=v_2^{\gamma}\left(\Pi+\frac{2\sigma^{\alpha\gamma}}{R_n}+\left.\frac{\partial\sigma^{\alpha\gamma}}{\partial R_n}\right|_{h=R-R_n}\right)$$
$$+\left(v_2^{\gamma}-v_2^{\alpha}\right)\left(\frac{2\sigma^{\alpha\beta}}{R}-\frac{R_n^2}{R^2}\Pi\right).$$

The equilibrium concentration x entering equation,

$$k_B T\ In\frac{a(x)}{a_{\infty}}=v_2^{\gamma}\left(\Pi+\frac{2\sigma^{\alpha\gamma}}{R_n}+\left.\frac{\partial\sigma^{\alpha\gamma}}{\partial R_n}\right|_{h=R-R_n}\right)+\left(v_2^{\gamma}-v_2^{\alpha}\right)\left(\frac{2\sigma^{\alpha\beta}}{R}-\frac{R_n^2}{R^2}\Pi\right).$$

refers not to the film itself, but to the film mother phase. It would coincide with the solubility x_{R_n} of the nanoparticle of radius R_n if the nanoparticle was immersed in the bulk phase α. The real solubility x^f of the nanoparticle matter in the film is different from x and x_{R_n} because the distribution of the solute in the film is inhomogeneous. Evidently, the real distribution of the solute in the film can be represented as the bulk concentration in phase α and adsorptions at the film boundaries. The adsorbed matter at the core surface is indistinguishable from the core itself and can easily be taken into account by the choice of radius R_n. As far as the adsorption at the film boundary with the gas phase is concerned, this adsorption is negligible in the typical case of surface inactive matter of the core (for instance, NaCl) that usually constitutes soluble condensation nuclei. Thus, the average solute concentration in the film and the bulk concentration in phase α can be considered to be almost equal at the same temperature and chemical potentials, and we may call the equilibrium concentration x the solubility of the solid core in the film.

Evidently,

equation $k_B T \; In \frac{a(x)}{a_\infty} = v_2^\gamma \left(\Pi + \frac{2\sigma^{\alpha\gamma}}{R_n} + \left. \frac{\partial \sigma^{\alpha\gamma}}{\partial R_n} \right|_{h=R-R_n} \right) + \left(v_2^\gamma - v_2^\alpha \right) \left(\frac{2\sigma^{\alpha\beta}}{R} - \frac{R_n^2}{R^2} \Pi \right)$ represents a more general form of th Ostwald–Freundlich equation $k_B T \; \text{In} = \frac{x_R}{x_\infty} = \frac{2\sigma^{\alpha\gamma} v_2^\gamma}{R}$ for the problem considered. Let us introduce some simplifications into the problem. Below we will consider only dilute ideal solutions and neglect the dependence of surface tensions $\sigma^{\alpha\beta}$ and $\sigma^{\alpha\gamma}$, partial volumes v_1^α, v_2^α, and the disjoining pressure Π on solute concentration.

Equations,

$$\mu_{1,0}^\alpha \left(p^\beta \right) + \frac{2\sigma^{\alpha\beta} v_1^\alpha}{R} - v_1^\alpha \left[\frac{R_n^2}{R^2} \Pi + \pi(x) \right]$$

and

$$k_B T \; In \frac{a(x)}{a_\infty} = v_2^\gamma \left(\Pi + \frac{2\sigma^{\alpha\gamma}}{R_n} + \left. \frac{\partial \sigma^{\alpha\gamma}}{\partial R_n} \right|_{h=R-R_n} \right) + \left(v_2^\gamma - v_2^\alpha \right) \left(\frac{2\sigma^{\alpha\beta}}{R} - \frac{R_n^2}{R^2} \Pi \right)$$

They can therewith be rewritten at as,

$$\mu_1 - \mu_{1,0}^\alpha \left(p^\beta \right) = \frac{2\sigma^{\alpha\beta} v_1^\alpha}{R} - v_1^\alpha \frac{R_n^2}{R^2} \Pi - k_B T x,$$

$$k_B T \; \text{In} \; \frac{x}{x_\infty} = \frac{2\sigma^{\alpha\beta} \left(v_2^\gamma - v_2^\alpha \right)}{R} + \left(\frac{2\sigma^{\alpha\gamma}}{R_n} + \left. \frac{\partial \sigma^{\alpha\gamma}}{\partial R_n} \right|_{h=R-R_n} \right) v_2^\gamma$$

$$-\left(v_2^\gamma - v_2^\alpha \right) \frac{R_n^2}{R^2} \Pi + v_2^\gamma \Pi.$$

In the particular case of a flat film, it follows from equation $\mu_1 - \mu_{1,0}^{\alpha}\left(p^{\beta}\right) = \frac{2\sigma^{\alpha\beta} v_1^{\alpha}}{R} - v_1^{\alpha}\frac{R_n^2}{R^2}\Pi - k_B Tx,$

and $k_B T \text{ In } \frac{x}{x_\infty} = \frac{2\sigma^{\alpha\beta}\left(v_2^{\gamma} - v_2^{\alpha}\right)}{R} + \left(\frac{2\sigma^{\alpha\gamma}}{R_n} + \left.\frac{\partial\sigma^{\alpha\gamma}}{\partial R_n}\right|_{h=R-R_n}\right) v_2^{\gamma} - \left(v_2^{\gamma} - v_2^{\alpha}\right)\frac{R_n^2}{R^2}\Pi + v_2^{\gamma}\Pi,$ as $R_n \to \infty$

and $R \to \infty$

$$\mu_1 - \mu_{1,\infty}^{\alpha} = v_1^{\alpha}\,\Pi(h) - k_B T_x,$$

$$k_B T \text{ In} = \frac{x}{x_\infty} = v_2^{\alpha}\,\Pi(h),$$

Where h is the flat film thickness.

The surface tensions $\sigma^{\alpha\gamma}$ and $\sigma^{\alpha\beta}$ at the boundaries of the film are related to the disjoining pressure Π .This cannot be neglected for a thin film in the general case. The thermodynamics of flat thin films gives the following relation:

$$\frac{\partial\left(\sigma^{\alpha\beta} + \sigma^{\alpha\gamma}\right)}{\partial h} = -\Pi(h).$$

Recognizing that the disjoining pressure is determined by the normal component of the pressure tensor at the internal boundary of the film and assuming the main contribution to equation, $\frac{\partial\left(\sigma^{\alpha\beta} + \sigma^{\alpha\gamma}\right)}{\partial h} = -\Pi(h)$ to be given by the surface tension $\sigma^{\alpha\gamma}$ at the same boundary, we can approximately replace the surface tension $\sigma^{\alpha\beta}$ at the external boundary of the film by its macroscopic value $\sigma_\infty^{\alpha\beta}$. It allows us to rewrite equation, $\frac{\partial\left(\sigma^{\alpha\beta} + \sigma^{\alpha\gamma}\right)}{\partial h} = -\Pi(h)$ in the form $\frac{\partial\sigma^{\alpha\gamma}}{\partial h} \simeq -\Pi(h)$.

Integrating equation, $\frac{\partial\sigma^{\alpha\gamma}}{\partial h} \simeq -\Pi(h)$ over thickness h at a fixed radius R_n.

We find,

$$\sigma^{\alpha\gamma} \simeq \sigma_{R_n}^{\alpha\gamma} + \int_{R-Rn}^{\infty}\Pi(h)\,dh,$$

where the surface tension $\sigma_{R_n}^{\alpha\gamma}$ corresponds, as before, to the solid core boundary with bulk phase α. Using equation, $\sigma^{\alpha\gamma} \simeq \sigma_{R_n}^{\alpha\gamma} + \int_{R-Rn}^{\infty}\Pi(h)\,dh,$ and equality $\sigma^{\alpha\beta} = \sigma_\infty^{\alpha\beta}$ equation,

$$\mu_1 - \mu_{1,0}^{\alpha}\left(p^{\beta}\right) = \frac{2\sigma^{\alpha\beta} v_1^{\alpha}}{R} - v_1^{\alpha}\frac{R_n^2}{R^2}\Pi - k_B Tx,$$

and

$$k_B T \text{ In } \frac{x}{x_\infty} = \frac{2\sigma^{\alpha\beta}\left(v_2^{\gamma} - v_2^{\alpha}\right)}{R} + \left(\frac{2\sigma^{\alpha\gamma}}{R_n} + \left.\frac{\partial\sigma^{\alpha\gamma}}{\partial R_n}\right|_{h=R-R_n}\right) v_2^{\gamma} - \left(v_2^{\gamma} - v_2^{\alpha}\right)\frac{R_n^2}{R^2}\Pi + v_2^{\gamma}\Pi$$

gives,

$$\mu_1 - \mu_{1,0}^{\alpha}\left(p^{\beta}\right) = \frac{2\sigma_{\infty}^{\alpha\beta} v_1^{\alpha}}{R} - v_1^{\alpha}\frac{R_n^2}{R^2}\Pi - k_B T x,$$

$$k\ T\ In\frac{}{x_{\infty}} = \frac{\left(v_2^{\gamma}\quad v_2^{\alpha}\right)\ {}_{\infty}^{\alpha\beta}}{R} + \frac{v^{\gamma}\sigma_{R_n}^{\alpha\gamma}}{R_n} + v\ \frac{d\sigma_{R_n}^{\alpha\gamma}}{dR_n}$$

$$-\left(v_2^{\gamma} - v_2^{\alpha}\right)\frac{}{R}\Pi + v_2^{\gamma}\Pi + \frac{}{R}\int_{R\ \ R}^{\infty}\Pi(h)\,dh$$

It follows from the condition of equilibrium of the film with the gas phase that $\mu_1 = \mu_1^{\beta}$ at $p_1^{\beta} = p_{1,R}^{\beta}$. We have already noted that $\mu_{1,0}^{\alpha}\left(p^{\beta}\right) \simeq \mu_{1,\infty}$ and $\mu_{1,\infty} = \mu_1^{\beta}\left(p_{1,\infty}^{\beta}\right)$, where $p_{1,\infty}^{\beta}$ is the saturated partial pressure of condensate vapor at a flat interface between pure liquid condensate and vapor. Thus, in the approximation of ideality of phase β, the difference $\mu_1 - \mu_{1,0}^{\alpha}\left(p^{\beta}\right)$ can be expressed as

$$\mu_1 - \mu_{1,0}^{\alpha}\left(p^{\beta}\right) \simeq k_B T\ In\frac{p_{1,R}^{\beta}}{p_{1,\infty}^{\beta}}.$$

Substituting equation, $\mu_1 - \mu_{1,0}^{\alpha}\left(p^{\beta}\right) \simeq k_B T\ In\frac{p_{1,R}^{\beta}}{p_{1,\infty}^{\beta}}.$

in the left-hand side of equation,

$$\mu_1 - \mu_{1,0}^{\alpha}\left(p^{\beta}\right) = \frac{2\sigma_{\infty}^{\alpha\beta} v_1^{\alpha}}{R} - v_1^{\alpha}\frac{R_n^2}{R^2}\Pi - k_B T x,$$

gives

$$k_B T\ \text{In}\ \frac{p_{1,R}^{\beta}}{p_{1,\infty}^{\beta}} = \frac{2\sigma^{\alpha\beta} v_1^{\alpha}}{R} - v_1^{\alpha}\frac{R_n^2}{R^2}\Pi - k_B T x.$$

Jointly with equation,

$$k_B T\ In\frac{x}{x_{\infty}} = \frac{2\left(v_2^{\gamma} - v_2^{\alpha}\right)\sigma_{\infty}^{\alpha\beta}}{R} + \frac{2v_2^{\gamma}\sigma_{R_n}^{\alpha\gamma}}{R_n} + v_2^{\gamma}\frac{d\sigma_{R_n}^{\alpha\gamma}}{dR_n} - \left(v_2^{\gamma} - v_2^{\alpha}\right)\frac{R_n^2}{R^2}\Pi + v_2^{\gamma}\Pi + \frac{2v_2^{\gamma}}{R_n}\int_{R-R_n}^{\infty}\Pi(h)\,dh.,$$

equation, $k_B T\ \text{In}\ \frac{p_{1,R}^{\beta}}{p_{1,\infty}^{\beta}} = \frac{2\sigma^{\alpha\beta} v_1^{\alpha}}{R} - v_1^{\alpha}\frac{R_n^2}{R^2}\Pi - k_B T x.$ determines the dependence of the pressure of the condensate vapor saturated over the spherical film of solution, on radii R_n and R of internal and external boundaries of the film. Thus, equation, $k_B T\ \text{In}\ \frac{p_{1,R}^{\beta}}{p_{1,\infty}^{\beta}} = \frac{2\sigma^{\alpha\beta} v_1^{\alpha}}{R} - v_1^{\alpha}\frac{R_n^2}{R^2}\Pi - k_B T x$is a generalization of the Gibbs–Kelvin–Köhler equation for the condensate vapor pressure saturated over a droplet with a spherical solid core partially dissolved in the droplet. With the choice of radii Rn and R of the internal and external film boundaries as independent variables of the film state, with specified isotherm of the disjoining pressure as a function of the film thickness and a known dependence of $\sigma_{R_n}^{\alpha\gamma}$ on Rn, equation,

$$k_B T \; In \frac{x}{x_\infty} = \frac{2\left(v_2^\gamma - v_2^\alpha\right)\sigma_\infty^{\alpha\beta}}{R} + \frac{2v_2^\gamma \sigma_{R_n}^{\alpha\gamma}}{R_n} + v_2^\gamma \frac{d\sigma_{R_n}^{\alpha\gamma}}{dR_n} - \left(v_2^\gamma - v_2^\alpha\right)\frac{R_n^2}{R^2}\Pi + v_2^\gamma \Pi + \frac{2v_2^\gamma}{R_n}\int_{R-R_n}^{\infty}\Pi(h)\,dh$$

first allows one to find the core solubility x, and then, with the aid of equation,

$$\mu_1 - \mu_{1,0}^\alpha\left(p^\beta\right) = \frac{2\sigma_\infty^{\alpha\beta} v_1^\alpha}{R} - v_1^\alpha \frac{R_n^2}{R^2}\Pi - k_B T x,$$

and

$$k_B T \; \text{In} \; \frac{p_{1,R}^\beta}{p_{1,\infty}^\beta} = \frac{2\sigma^{\alpha\beta} v_1^\alpha}{R} - v_1^\alpha \frac{R_n^2}{R^2}\Pi - k_B T x.,$$

also to determine the chemical potential of the condensate in the film and the pressure of the condensate vapor saturated over the film. Doing in this way solves the problem posed in Introduction.

Note now that radii R_n and R at the internal and external film boundaries may be determined as functions of the number N_1^α of condensate molecules and the number N_2^α of dissolved core molecules in the solution film (one can use the total number N_2 of molecules of the core component instead of N_2^α). As it follows from the conditions of material balance and incompressibility in the core and the solution film, we have,

$$\frac{4}{3}\pi R_n^3 = v_2^\gamma\left(N_2 - N_2^\alpha\right),$$

$$\frac{4}{3}\pi R^3 = \left(N_2 - N_2^\alpha\right)v_2^\gamma + N_1^\alpha v_1^\alpha + N_2^\alpha v_2^\alpha.$$

Substituting equation, $\frac{4}{3}\pi R_n^3 = v_2^\gamma\left(N_2 - N_2^\alpha\right)$, and $\frac{4}{3}\pi R^3 = \left(N_2 - N_2^\alpha\right)v_2^\gamma + N_1^\alpha v_1^\alpha + N_2^\alpha v_2^\alpha$. in equation,

$$k_B T \; In \frac{x}{x_\infty} = \frac{2\left(v_2^\gamma - v_2^\alpha\right)\sigma_\infty^{\alpha\beta}}{R} + \frac{2v_2^\gamma \sigma_{R_n}^{\alpha\gamma}}{R_n} + v_2^\gamma \frac{d\sigma_{R_n}^{\alpha\gamma}}{dR_n} - \left(v_2^\gamma - v_2^\alpha\right)\frac{R_n^2}{R^2}\Pi + v_2^\gamma \Pi + \frac{2v_2^\gamma}{R_n}\int_{R-R_n}^{\infty}\Pi(h)\,dh.$$

Transforms equation,

$$k_B T \; In \frac{x}{x_\infty} = \frac{2\left(v_2^\gamma - v_2^\alpha\right)\sigma_\infty^{\alpha\beta}}{R} + \frac{2v_2^\gamma \sigma_{R_n}^{\alpha\gamma}}{R_n} + v_2^\gamma \frac{d\sigma_{R_n}^{\alpha\gamma}}{dR_n} - \left(v_2^\gamma - v_2^\alpha\right)\frac{R_n^2}{R^2}\Pi + v_2^\gamma \Pi + \frac{2v_2^\gamma}{R_n}\int_{R-R_n}^{\infty}\Pi(h)\,dh$$

(with account of equality $x \equiv N_2^\alpha / N_1^\alpha$) into a transcendental equation for N_1^α and N_2^α.Considering number N_2 as a parameter, this equation can be solved at constant surface tensions and the assumption that $N_1^\alpha = N_1$, where N_1 is the total number of condensate molecules in the film. The function $N_1^\alpha(N_1)$ obtained in this way allows one to determine R_n and R with the use of equation,

$$\frac{4}{3}\pi R_n^3 = v_2^\gamma\left(N_2 - N_2^\alpha\right), \text{ and } \frac{4}{3}\pi R^3 = \left(N_2 - N_2^\alpha\right)v_2^\gamma + N_1^\alpha v_1^\alpha + N_2^\alpha v_2^\alpha,$$

i.e., to find the function $R_n(R)$ at a specified N_2. Substituting this function in equation,

$$k_B T \text{ In } \frac{p_{1,R}^\beta}{p_{1,\infty}^\beta} = \frac{2\sigma^{\alpha\beta} v_1^\alpha}{R} - v_1^\alpha \frac{R_n^2}{R^2}\Pi - k_B T x$$

establishes a relation between the pressure of condensate vapor saturated above the liquid film on a partially dissolved solid core and the external radius of the film at a specified initial (i.e., before the film formation) size of the core. Such a relation can be observed in direct experiment with soluble solid nanoparticles in the undersaturated solvent gas environment.

Pitzer Equations

Pitzer equations are important for the understanding of the behaviour of ions dissolved in natural waters such as rivers, lakes and sea-water. They were first described by physical chemist Kenneth Pitzer. The parameters of the Pitzer equations are linear combinations of parameters, of a virial expansion of the excess Gibbs free energy, which characterise interactions amongst ions and solvent. The derivation is thermodynamically rigorous at a given level of expansion. The parameters may be derived from various experimental data such as the osmotic coefficient, mixed ion activity coefficients, and salt solubility. They can be used to calculate mixed ion activity coefficients and water activities in solutions of high ionic strength for which the Debye–Hückel theory is no longer adequate. They are more rigorous than the equations of specific ion interaction theory (SIT theory), but Pitzer parameters are more difficult to determine experimentally than SIT parameters.

A starting point for the development can be taken as the virial equation of state for a gas.

$$PV = RT + BP + CP^2 + DP^3 \ldots$$

where P is the pressure, V is the volume, T is the temperature and B, C, D... are known as virial coefficients. The first term on the right-hand side is for an ideal gas. The remaining terms quantify the departure from the ideal gas law with changing pressure, P. It can be shown by statistical mechanics that the second virial coefficient arises from the intermolecular forces between *pairs* of molecules, the third virial coefficient involves interactions between three molecules, etc. This theory was developed by McMillan and Mayer.

Solutions of uncharged molecules can be treated by a modification of the McMillan-Mayer theory. However, when a solution contains electrolytes, electrostatic interactions must also be taken into account. The Debye-Hückel theory was based on the assumption that each ion was surrounded by a spherical "cloud" or ionic atmosphere made up of ions of the opposite charge. Expressions were derived for the variation of single-ion activity coefficients as a function of ionic strength. This theory was very successful for dilute solutions of 1:1 electrolytes and, as discussed below, the Debye-Hückel expressions are still valid at sufficiently low concentrations. The values calculated with Debye-Hückel theory diverge more and more from observed values as the concentrations and/or ionic charges increases. Moreover, Debye-Hückel theory takes no account of the specific properties of ions such as size or shape.

Brønsted had independently proposed an empirical equation,

$$\ln \gamma = -\alpha m^{1/2} - 2\beta m$$
$$1 - \varphi = (\alpha / 3) m^{1/2} + \beta m$$

In which the activity coefficient depended not only on ionic strength, but also on the concentration, m, of the specific ion through the parameter β. This is the basis of SIT theory. It was further developed by Guggenheim. Scatchard extended the theory to allow the interaction coefficients to vary with ionic strength. Note that the second form of Brønsted's equation is an expression for the osmotic coefficient. Measurement of osmotic coefficients provides one means for determining mean activity coefficients.

The Pitzer Parameters

The exposition begins with a virial expansion of the excess Gibbs free energy:

$$\frac{G^{ex}}{W_w RT} = f(I) + \sum_i \sum_j b_i b_j \lambda_{ij}(I) + \sum_i \sum_j \sum_k b_i b_j b_k \mu_{ijk} + \cdots$$

W_w is the mass of the water in kilograms, b_i, b_j ... are the molalities of the ions and I is the ionic strength. The first term, *f(I)* represents the Debye-Hückel limiting law. The quantities $\lambda_{ij}(I)$ represent the short-range interactions in the presence of solvent between solute particles i and j. This binary interaction parameter or second virial coefficient depends on ionic strength, on the particular species i and j and the temperature and pressure. The quantities μ_{ijk} represent the interactions between three particles. Higher terms may also be included in the virial expansion.

Next, the free energy is expressed as the sum of chemical potentials, or partial molal free energy,

$$G = \sum_i \mu_i \cdot N_i = \sum_i \left(\mu_i^0 + RT \ln b_i \gamma_i \right) \cdot N_i$$

and an expression for the activity coefficient is obtained by differentiating the virial expansion with respect to a molality b.

$$\ln \gamma_i = \frac{\partial\left(\frac{G^{ex}}{W_w RT}\right)}{\partial b_i} = \frac{z_i^2}{2} f' + 2\sum_j \lambda_{ij} b_j + \frac{z_i^2}{2} \sum_j \sum_k \lambda'_{jk} b_j b_k + 3\sum_j \sum_k \mu_{ijk} b_j b_k + \cdots$$

$$\phi-1=\left(\sum_i b_i\right)^{-1}\left[If'-f+\sum_i\sum_j\left(\lambda_{ij}+I\lambda'_{ij}\right)b_ib_j+2\sum_i\sum_j\sum_k\mu_{ijk}b_ib_jb_k+\cdots\right]$$

For a simple electrolyte M_pX_q, at a concentration m, made up of ions M^{z+} and X^{z-}, the parameters f^ϕ, B^ϕ_{MX} and C^ϕ_{MX} are defined as:

$$f^\phi=\frac{f'-\dfrac{f}{I}}{2}$$

$$B_{MX}=\lambda_{MX}+I\lambda'_{MX}+\left(\frac{p}{2q}\right)\left(\lambda_{MM}+I\lambda'_{MM}\right)+\left(\frac{q}{2p}\right)\left(\lambda_{XX}+I\lambda'_{XX}\right)$$

$$C^\phi_{MX}=\left[\frac{3}{\sqrt{pq}}\right]\left(p\mu_{MMX}+q\mu_{MXX}\right).$$

The term f^ϕ is essentially the Debye-Hückel term. Terms involving μ_{MMM} and μ_{XXX} are not included as interactions between three ions of the same charge are unlikely to occur except in very concentrated solutions.

The B parameter was found empirically to show an ionic strength dependence (in the absence of ion-pairing) which could be expressed as:

$$B^\phi_{MX}=\beta^{(0)}_{MX}+\beta^{(1)}_{MX}e^{-\alpha\sqrt{I}}.$$

With these definitions, the expression for the osmotic coefficient becomes:

$$\phi-1=|z^+z^-|\,f^\phi+b\left(\frac{2pq}{p+q}\right)B^\phi_{MX}+m^2\left[2\frac{(pq)^{3/2}}{p+q}\right]C^\phi_{MX}.$$

A similar expression is obtained for the mean activity coefficient.

$$\ln\gamma_\pm=\frac{p\ln\gamma_M+q\ln\gamma_X}{p+q}$$

$$\ln\gamma_\pm=|z^+z^-|\,f^\gamma+m\left(\frac{2pq}{p+q}\right)B^\gamma_{MX}+m^2\left[2\frac{(pq)^{3/2}}{p+q}\right]C^\gamma_{MX}$$

These equations were applied to an extensive range of experimental data at 25 °C with excellent agreement to about 6 mol kg^{-1} for various types of electrolyte. The treatment can be extended to mixed electrolytes and to include association equilibria. Values for the parameters $\beta^{(0)}$, $\beta^{(1)}$ and C for inorganic and organic acids, bases and salts have been tabulated. Temperature and pressure variation is also discussed.

One area of application of Pitzer parameters is to describe the ionic strength variation of equilibrium constants measured as concentration quotients. Both SIT and Pitzer parameters have been used in this context, For example, both sets of parameters were calculated for some uranium complexes and were found to account equally well for the ionic strength dependence of the stability constants.

Pitzer parameters and SIT theory have been extensively compared. There are more parameters in the Pitzer equations than in the SIT equations. Because of this the Pitzer equations provide for more precise modelling of mean activity coefficient data and equilibrium constants. However, the determination of the greater number of Pitzer parameters means that they are more difficult to determine.

Compilation of Pitzer Parameters

Besides the set of parameters obtained by Pitzer et al. in the 1970s. Kim and Frederick published the Pitzer parameters for 304 single salts in aqueous solutions at 298.15 K, extended the model to the concentration range up to the saturation point. Those parameters are widely used, however, many complex electrolytes including ones with organic anions or cations, which are very significant in some related fields, were not summarized in their paper.

For some complex electrolytes, Ge et al. obtained the new set of Pitzer parameters using up-to-date measured or critically reviewed osmotic coefficient or activity coefficient data.

Comparable TCPC Model

Besides the well-known Pitzer-like equations, there is a simple and easy-to-use semi-empirical model, which is called the three-characteristic-parameter correlation (TCPC) model. It was first proposed by Lin et al. It is a combination of the Pitzer long-range interaction and short-range solvation effect:

$$\ln \gamma = \ln \gamma^{\text{PDH}} + \ln \gamma^{\text{SV}}$$

Ge et al. modified this model, and obtained the TCPC parameters for a larger number of single salt aqueous solutions. This model was also extended for a number of electrolytes dissolved in methanol, ethanol, 2-propanol, and so on. Temperature dependent parameters for a number of common single salts were also compiled, available at.

The performance of the TCPC model in correlation with the measured activity coefficient or osmotic coefficients is found to be comparable with Pitzer-like models.

Clausius Clapeyron Equation

Clausius–Clapeyron equation is the differential equation relating pressure of a substance to temperature in a system in which two phases of the substance are in equilibrium.

Two general expressions are:

$$\frac{dp}{dT} = \frac{\delta s}{\delta v} = \frac{L}{T\delta v},$$

where p is the pressure, T the temperature, δs the difference in specific entropy between the phases, δv the difference in specific volume between the two phases, and L the latent heat of the phase change. The form most familiar in meteorology, related to the phase change between water vapor and liquid water, is obtained after some approximations as,

$$\frac{1}{e_s}\frac{de_s}{dT} = \frac{L_v}{R_v T^2}$$

where e_s is the saturation vapor pressure of water, L_v the latent heat of vaporization, and R_v the gas constant for water vapor. A similar relation for the saturation vapor pressure in contact with an ice surface is obtained by replacing the latent heat of vaporization by that of sublimation. These equations may be integrated to obtain explicit relationships between e_s and T, given known values at some point. The most empirically accurate relationships differ slightly from results so obtained. An expression believed accurate to 0.3% for -35 °C < T < 35 °C is given by Bolton as,

$$e_s(T) = 0.6112 \exp\left(\frac{17.67T}{T + 243.5}\right),$$

where T is temperature in °C and vapor pressure is in kPa.

Derivations

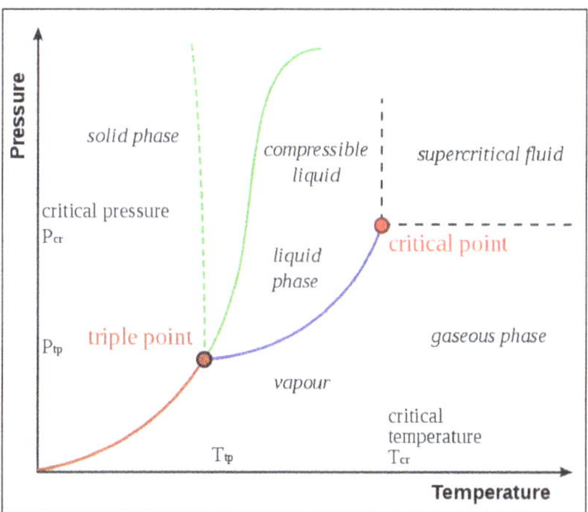

A typical phase diagram. The dotted green line gives the anomalous behavior of water. The Clausius–Clapeyron relation can be used to find the relationship between pressure and temperature along phase boundaries.

Derivation from State Postulate

Using the state postulate, take the specific entropy s for a homogeneous substance to be a function of specific volume v and temperature T,

$$\mathrm{d}s = \left(\frac{\partial s}{\partial v}\right)_T \mathrm{d}v + \left(\frac{\partial s}{\partial T}\right)_v \mathrm{d}T.$$

The Clausius–Clapeyron relation characterizes behavior of a closed system during a phase change, during which temperature and pressure are constant by definition. Therefore,

$$ds = \left(\frac{\partial s}{\partial v}\right)_T dv.$$

Using the appropriate Maxwell relation gives:

$$ds = \left(\frac{\partial P}{\partial T}\right)_v dv$$

where P is the pressure. Since pressure and temperature are constant, by definition the derivative of pressure with respect to temperature does not change. Therefore, the partial derivative of specific entropy may be changed into a total derivative:

$$ds = \frac{dP}{dT} dv$$

and the total derivative of pressure with respect to temperature may be factored out when integrating from an initial phase α to a final phase β to obtain:

$$\frac{dP}{dT} = \frac{\Delta s}{\Delta v}$$

where $\Delta s \equiv s_\beta - s_\alpha$ and $\Delta v \equiv v_\beta - v_\alpha$ are respectively the change in specific entropy and specific volume. Given that a phase change is an internally reversible process, and that our system is closed, the first law of thermodynamics holds:

$$du = \delta q + \delta w = T ds - P dv$$

where u is the internal energy of the system. Given constant pressure and temperature (during a phase change) and the definition of specific enthalpy h, we obtain:

$$dh = T ds + v\,dP$$

$$dh = T ds$$

$$ds = \frac{dh}{T}$$

Given constant pressure and temperature (during a phase change), we obtain:

$$\Delta s = \frac{\Delta h}{T}$$

Substituting the definition of specific latent heat $L = \Delta h$ gives:

$$\Delta s = \frac{L}{T}$$

Substituting this result into the pressure derivative given above,

$$\left(dP / dT = \Delta \mathrm{s} / \Delta \mathrm{v}\right),$$

we obtain:

$$\frac{dP}{dT} = \frac{L}{T\,\Delta v}.$$

This result (also known as the Clapeyron equation) equates the slope of the tangent to the coexistence curve dP/dT , at any given point on the curve, to the function $L/(T\,\Delta v)$ of the specific latent heat L , the temperature T, and the change in specific volume Δv.

Derivation from Gibbs–Duhem Relation

Suppose two phases, α and β, are in contact and at equilibrium with each other. Their chemical potentials are related by:

$$\mu_\alpha = \mu_\beta.$$

Furthermore, along the coexistence curve,

$$d\mu_\alpha = d\mu_\beta.$$

One may therefore use the Gibbs–Duhem relation:

$$d\mu = M(-s\,dT + v\,dP)$$

(Where s is the specific entropy, v is the specific volume, and M is the molar mass) to obtain:

$$-(s_\beta - s_\alpha)\,dT + (v_\beta - v_\alpha)\,dP = 0$$

Rearrangement gives:

$$\frac{dP}{dT} = \frac{s_\beta - s_\alpha}{v_\beta - v_\alpha} = \frac{\Delta s}{\Delta v}$$

From which the derivation of the Clapeyron equation continues.

Ideal Gas Approximation at Low Temperatures

When the phase transition of a substance is between a gas phase and a condensed phase (liquid or solid), and occurs at temperatures much lower than the critical temperature of that substance, the specific volume of the gas phase v_g greatly exceeds that of the condensed phase v_c . Therefore, one may approximate:

$$\Delta v = v_g\left(1 - \tfrac{v_c}{v_g}\right) \approx v_g$$

At low temperatures. If pressure is also low, the gas may be approximated by the ideal gas law, so that:

$$v_g = \frac{RT}{P}$$

Where P is the pressure, R is the specific gas constant, and T is the temperature. Substituting into the Clapeyron equation:

$$\frac{dP}{dT} = \frac{L}{T\,\Delta v}$$

We can obtain the Clausius–Clapeyron equation:

$$\frac{dP}{dT} = \frac{PL}{T^2 R}$$

For low temperatures and pressures, where L is the specific latent heat of the substance.

Let (P_1, T_1) and (P_2, T_2) be any two points along the coexistence curve between two phases α and β. In general, L varies between any two such points, as a function of temperature. But if L is constant,

$$\frac{dP}{P} = \frac{L}{R}\frac{dT}{T^2},$$

$$\int_{P_1}^{P_2} \frac{dP}{P} = \frac{L}{R}\int_{T_1}^{T_2} \frac{dT}{T^2}$$

$$\ln P \Big|_{P=P_1}^{P_2} = -\frac{L}{R}\cdot\frac{1}{T}\Big|_{T=T_1}^{T_2}$$

or

$$\ln \frac{P_2}{P_1} = -\frac{L}{R}\left(\frac{1}{T_2} - \frac{1}{T_1}\right)$$

These last equations are useful because they relate equilibrium or saturation vapor pressure and temperature to the latent heat of the phase change, without requiring specific volume data.

Applications

Chemistry and Chemical Engineering

For transitions between a gas and a condensed phase with the approximations, the expression may be rewritten as:

$$\ln P = -\frac{L}{R}\left(\frac{1}{T}\right) + c$$

Where c is a constant. For a liquid-gas transition, L is the specific latent heat (or specific enthalpy) of vaporization; for a solid-gas transition, L is the specific latent heat of sublimation. If the latent heat is known, then knowledge of one point on the coexistence curve determines the rest of the curve. Conversely, the relationship between ln P and 1 / T is linear, and so linear regression is used to estimate the latent heat.

Meteorology and Climatology

Atmospheric water vapor drives many important meteorologic phenomena (notably precipitation), motivating interest in its dynamics. The Clausius–Clapeyron equation for water vapor under typical atmospheric conditions (near standard temperature and pressure) is:

$$\frac{de_s}{dT} = \frac{L_v(T)e_s}{R_v T^2}$$

where:

- e_s is saturation vapor pressure,
- T is temperature,
- L_v is the specific latent heat of evaporation of water,
- R_v is the gas constant of water vapour.

The temperature dependence of the latent heat $L_v(T)$, and therefore of the saturation vapor pressure $e_s(T)$, cannot be neglected in this application. Fortunately, the August–Roche–Magnus formula provides a very good approximation, using pressure in hPa and temperature in Celsius:

$$e_s(T) = 6.1094 \exp\left(\frac{17.625T}{T + 243.04}\right)$$

(This is also sometimes called the Magnus or Magnus–Tetens approximation, though this attribution is historically inaccurate).

Under typical atmospheric conditions, the denominator of the exponent depends weakly on T (for which the unit is Celsius). Therefore, the August–Roche–Magnus equation implies that saturation water vapor pressure changes approximately exponentially with temperature under typical atmospheric conditions, and hence the water-holding capacity of the atmosphere increases by about 7% for every 1 °C rise in temperature.

One of the uses of this equation is to determine if a phase transition will occur in a given situation. Consider the question of how much pressure is needed to melt ice at a temperature ΔT below 0 °C. Note that water is unusual in that its change in volume upon melting is negative. We can assume

$$\Delta P = \frac{}{T \quad v} \Delta T$$

and substituting in:

$L = 3.34 \times 10^5$ J / kg (latent heat of fusion for water),

$T = 273$ K (absolute temperature),

$\Delta v = -9.05 \times 10^{-5}$ m^3 / kg (change in specific volume from solid to liquid),

We obtain:

$$\frac{\Delta P}{\Delta T} = -135 \text{ MPa}/\text{K}$$

To provide a rough example of how much pressure this is, to melt ice at −7 °C (the temperature many ice skating rinks are set at) would require balancing a small car (mass = 1000 kg) on a thimble (area = 1 cm²).

Second Derivative

While the Clausius–Clapeyron relation gives the slope of the coexistence curve, it does not provide any information about its curvature or second derivative. The second derivative of the coexistence curve of phases 1 and 2 is given by,

$$\frac{\mathrm{d}^2 P}{\mathrm{d}T^2} = \frac{1}{v_2 - v_1}\left[\frac{c_{p2} - c_{p1}}{T} - 2(v_2\alpha_2 - v_1\alpha_1)\frac{\mathrm{d}P}{\mathrm{d}T}\right] +$$
$$\frac{1}{v_2 - v_1}\left[(v_2\kappa_{T2} - v_1\kappa_{T1})\left(\frac{\mathrm{d}P}{\mathrm{d}T}\right)^2\right],$$

where subscripts 1 and 2 denote the different phases, c_p is the specific heat capacity at constant pressure, $\alpha = (1/v)(\mathrm{d}v/\mathrm{d}T)_P$ is the thermal expansion coefficient, and $\kappa_T = -(1/v)(\mathrm{d}v/\mathrm{d}P)_T$ is the isothermal compressibility.

Antoine Equation

The Antoine equation is a mathematical expression (derived from the Clausius-Clapeyron relation) of the relation between the vapor pressure and the temperature of pure substances. The equation was first proposed by Ch. Antoine, a French researcher, in 1888. The basic form of the equation is:

$$\log P = A - \frac{B}{C + T}$$

and it can be transformed into this temperature-explicit form:

$$T = \frac{B}{A - \log P} - C$$

where:

P is the absolute vapor pressure of a substance,

T is the temperature of the substance,

A, *B* and *C* are substance-specific coefficients (i.e., constants or parameters), log is typically either $\log_{10}$ or $\log_e$.

A simpler form of the equation with only two coefficients is sometimes used:

$$\log P = A - \frac{B}{T}$$

Which can be transformed to:

$$T = \frac{B}{A - \log P}$$

Validity Ranges

The Antoine equation cannot be used for the entire vapor pressure range from the triple point to the critical point because it is not flexible enough. Therefore two sets of coefficients are commonly used: one set for vapor pressures at temperatures below the normal boiling point (NBP) and one set for vapor pressures at temperatures above the normal boiling point.

Example Sets of Coefficients

Table lists Antoine equation coefficients for water and for ethanol with each having two sets of coefficients: one for the temperature range below the normal boiling point (NBP) and one for the temperature range above the NBP. The temperature ranges are denoted by the indicated minimum and maximum temperatures.

	A	B	C	T minimum	T maximum
Water below the NBP	8.07131	1730.63	233.426	1	100
Water above the NBP	8.14019	1810.94	244.485	99	374
Ethanol below the NBP	8.20417	1642.89	230.300	-57	80
Ethanol below the NBP	7.68117	1332.04	199.200	77	243

The coefficients in table are for temperatures in °C and absolute pressures in mmHg when using log10 as the logarithmic function.

Example Calculations

The NBP of ethanol is 78.32 °C. Using the Antoine coefficient range for below the NBP from table, the ethanol vapor pressure at the NBP temperature is:

$$\log P_{10} = 8.20417 - \frac{1642.89}{230000 + 7832} = 2.8808$$

and

$$P = 10^{2.8808} = 760.02 \text{ mmHg}$$

Using the Antoine coefficient range for above the NBP from table, the ethanol vapor pressure at the NBP temperature is:

$$\log P_{10} = 7.68117 - \frac{1332.04}{199200 + 7832} = 2.8814$$

and

$$P = 10^{2.8814} = 760.98\,\text{mmHg}$$

(760 mmHg = 1.000 atm = typical atmospheric pressure at sea level)

This example shows the problem caused by using two different sets of coefficients. The two sets of coefficients give different results at the NBP temperature. This causes problems for computational techniques which rely on a continuous vapor pressure curve.

Two solutions are possible: the first approach uses a single Antoine parameter set over a larger temperature range and accepts the increased deviation between calculated and real vapor pressures. A variant of this single set approach is using a parameter set specially fitted for the desired temperature range. The second approach is to use the equations of Wagner or of the AIChE's Design Institute for Physical Properties (DIPPR).

Units

When using Antoine equation coefficients obtained from the technical literature, care must be taken to ascertain the units involved in such coefficients. Some of the coefficients provided in the technical literature are based on using $\log_{10}$ while others are based on using $\log_e$, and some are based on using the Celsius scale for temperatures while others are based on using the Kelvin scale. The pressure basis may be in various units other than mmHg. Occasionally, the coefficient B may be given as negative because the basic form of the Antoine equation has been rewritten with the minus sign changed to a plus sign.

It is relatively easy to convert the Antoine coefficients based on using Celsius scale temperatures to make them suitable for using Kelvin scale temperatures. All that is required is to subtract 273.15 from the C coefficient.

It is also relatively easy to convert coefficients based on using $\log_{10}$ and pressures in mmHg to make them suitable for using pressures in Pa. Since an absolute pressure of 101,325 Pa is equivalent to an absolute pressure of 760 mmHg, adding the $\log_{10}$ of (101,325 ÷ 760) to the A coefficient is all that is required:

$$A_{\text{Pa}} = A_{\text{mmHg}} + \log_{10}\frac{101325}{760} = A_{\text{mmHg}} + 2.124903$$

The Antoine coefficients for ethanol (below its NBP) based on using log10, mmHg and °C are:

A	B	C
8.20417	1642.89	230.300

When the coefficients in table are converted for using Kelvin scale temperatures and pressures in Pa, they are:

A	B	C
10.32907	1642.89	-42.85

Thus, for an ethanol NBP of 78.32 °C = 351 K and using the converted coefficients in table:

$$\log_{10} P = 10.32907 - \frac{1642.89}{35147 - 4285} = 5.00573$$

and

$$P = 10^{5.00573} = 101,328 \text{ Pa}$$

To make the coefficients in table (based on using $\log_{10}$) suitable for using $\log_e$, the A and B coefficients must be multiplied by $\log_e(10) = 2.302585$ to obtain:

A	B	C
23.7836	3782.89	-42.85

Now calculating the vapor pressure using the $\log_e$ yields:

$$\log_e P = 23.7836 - \frac{3782.89}{35147 - 4285} = 11.52616$$

and

$$P = e^{11.52616} = 101,332 \text{ Pa}$$

(The 0.006% difference in the calculated vapor pressures between using the log10 coefficients and the loge coefficients is due to the rounding off of numbers used in the calculations.)

Gibbs–Duhem Equation

The Gibbs free energy can be defined in two different ways once by subtracting off combinations of entropy S, enthalpy H and temperature T and other as a sum of chemical potentials and amounts of species. The fact that they are equal gives a new relation known as "Gibbs-Duhem Relation." The Gibbs-Duhem relation helps us to calculate relationships between quantities as a system which remains in equilibrium. One example is the Clausius-Clapeyron equation which states that two phases at equilibrium with each other having equaled amount of a given substance must have exactly the same free energy i.e. it relates equilibrium changes in pressure to changes in temperature as a function of material parameters.

Deriving the Gibbs-Duhem equation from thermodynamics state equations is very easy. The Gibbs free energy G in equilibrium can be expressed in terms of thermodynamics as:

$$dG = \mu_1\, dn_1 + n_1\, d\mu_1 + \mu_2\, dn_2 + n_2\, d\mu_2 \ldots\ldots\ldots\, \mu_j\, dn_j + n_j\, d\mu_j$$

$$= \left(\mu_1\, dn_1 + \mu_2\, dn_2 + \ldots\ldots\ldots\, \mu_j\, dn_j\right) + \left(n_1\, d\mu_1 + n_2\, d\mu_2 + \ldots\ldots\ldots\, n_j\, d\mu_j\right)$$

At constant temperature and pressure, the above equation can be written as:

$$n_1\, d\mu_1 + n_2\, d\mu_2 + \ldots\ldots\ldots\, n_j\, d\mu_j = 0$$

$$\Sigma\ n_i\, d\mu_i = 0$$

Because at constant temperature and pressure, $(\mu_1\, dn_1 + \mu_2\, dn_2 + \ldots\ldots\ldots\, \mu_j\, dn_j) = dG$

The equation $\Sigma\ n_i\, d\mu_i = 0$ is known as the Gibbs-Duhem equation.

Applications of Gibbs-Duhem Equation

- Gibbs-duhem equation is helpful in calculating partial molar quantity of a binary mixture by measuring the composition of the mixture which depends on the total molar quantity.
- Gibbs-duhem equation is helpful in calculating the partial vapor pressures by calculating the total vapor pressure. All these calculations require a curve-fitting procedure. Using tabulated experimental data the accuracy of the calculated quantities was found to be comparable to the accuracy of the original experimental data.

Gibbs–Helmholtz Equation

The name Gibbs-Helmholtz is usually associated with the equation:

$$H = G - T\left(\partial G / \partial T\right)_p$$

where H is enthalpy, G is Gibbs free energy and T temperature, with the implied assumption that the differentiation is to be carried out at constant composition. However, this is just one of a very large number of equations of similar form, which, at various times have been called Gibbs-Helmholtz equations. The equation given is, however, most appropriately called the Gibbs-Helmholtz equation since it was first recognized by both Gibbs and Helmholtz (independently).

The greatest utility of this equation lies in the presence of the temperature derivative of the Gibbs free energy. For example, the equilibrium constant in a chemical reaction is determined by the standard Free Energy change for the reaction. Thus, a knowledge of the temperature derivative of the free energy provides information relating the temperature derivative of the equilibrium constant. The equilibrium constant, K, may be written as:

$$\Delta g^\circ = -\tilde{R}T\ \text{In}\ K$$

where Δg^{o} is the standard Gibbs free energy change for the reaction at temperature T and $\tilde{R}$ is the universal gas constant.

Applying the Gibbs-Helmholtz equation to each component in the reaction gives:

$$\Delta h^{\circ} = \Delta g^{\circ} - T\left(\partial \Delta g^{\circ} / \partial T\right)_{p}$$

and hence,

$$(\mathrm{d}\ln \mathrm{K}) / \left(d[1/T]\right) = -\Delta h^{\circ} / \tilde{R}$$

Joule–Thomson Effect

The Joule-Thomson effect also known as Kelvin–Joule effect or Joule-Kelvin effect is the change in fluid's temperature as it flows from a higher pressure region to a lower pressure.

According to the thermodynamic principle, the Joule-kelvin effect can be explained best by considering a separate gas packet placed in the opposite flow of direction for restriction. For the gas packet to pass through, the upstream gas needs to perform some work to push through the packet. The work equals the volume of the packet multiplied by the times of upstream pressure:

$$W_1 = V_{packet\,1} \times P_1$$

As the packet goes through the restriction, it has to make some room by displacing a considerable amount of the downstream gas. It includes performing the work which equals to the product of packet volume and downstream pressure:

$$W_2 = V_{packet\,2} \times P_2$$

Due to the different effects of compressibility, the work performed upstream is not equal to the amount of work done downstream for real gases. Since depressuring is viewed as an adiabatic process, it reveals that any gas does not exchange work or heat by its surroundings, any change in internal energy has to follow the first law of thermodynamics:

$$U_2 - U_1 = W_1 - W_2$$

Gas molecules are subjected to repulsive and attractive forces, (Van der Waals forces) as they are in random motion. When the gas pressure is lowered, i.e, the average distance between the molecules increases, the attractive forces become dominant for many gases at ambient temperature which results in an elevation in potential energy:

Most of the real gases need more work downstream at ambient temperature, due to the effects of compressibility:

$$P_1 \times V_1 < P_2 \times V_2$$

The indicates that the internal energy decreases when the gas passes through the restriction.

It can be generalised that for many real gases, the temperature decreases during a reduction in pressure, but not true for every gas and condition. Depressuring is an isenthalpic process which reveals that enthalpy doesn't change. The temperature can either decrease or increase for any gas based on how the internal energy changes to maintain the enthalpy constant.

The *adiabatic* (no heat exchanged) expansion of a gas may be carried out in a number of ways. The change in temperature experienced by the gas during expansion depends not only on the initial and final pressure, but also on the manner in which the expansion is carried out.

- If the expansion process is reversible, meaning that the gas is in thermodynamic equilibrium at all times, it is called an *isentropic* expansion. In this scenario, the gas does positive work during the expansion, and its temperature decreases.
- In a free expansion, on the other hand, the gas does no work and absorbs no heat, so the internal energy is conserved. Expanded in this manner, the temperature of an ideal gas would remain constant, but the temperature of a real gas decreases, except at very high temperature.
- The method of expansion, in which a gas or liquid at pressure P_1 flows into a region of lower pressure P_2 without significant change in kinetic energy, is called the Joule–Thomson expansion. The expansion is inherently irreversible. During this expansion, enthalpy remains unchanged. Unlike a free expansion, work is done, causing a change in internal energy. Whether the internal energy increases or decreases is determined by whether work is done on or by the fluid; that is determined by the initial and final states of the expansion and the properties of the fluid.

The temperature change produced during a Joule–Thomson expansion is quantified by the Joule–Thomson coefficient, μ_{JT}. This coefficient may be either positive (corresponding to cooling) or negative (heating); the regions where each occurs for molecular nitrogen, N_2, are shown in the figure. Note that most conditions in the figure correspond to N_2 being a supercritical fluid, where it has some properties of a gas and some of a liquid, but can not be really described as being either. The coefficient is negative at both very high and very low temperatures; at very high pressure it is negative at all temperatures. The maximum inversion temperature (621 K for N_2) occurs as zero pressure is approached. For N_2 gas at low pressures, μ_{JT} is negative at high temperatures and positive at low temperatures. At temperatures below the gas-liquid coexistence curve, N_2 condenses to form a liquid and the coefficient again becomes negative. Thus, for N_2 gas below 621 K, a Joule–Thomson expansion can be used to cool the gas until liquid N_2 forms.

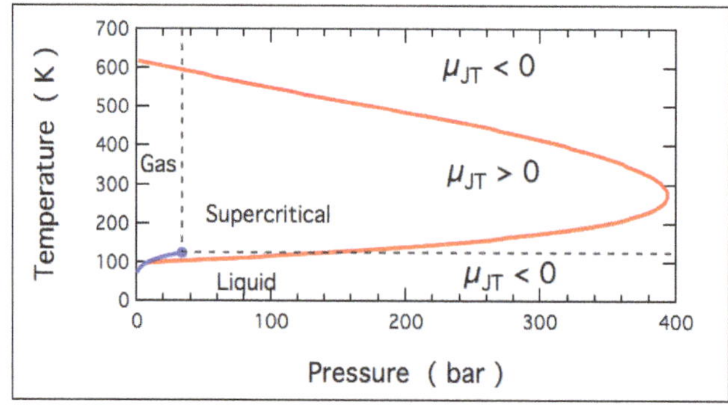

Sign of the Joule–Thomson coefficient, μ_{JT} for N_2. Within the region bounded by the red line, a Joule–Thomson expansion produces cooling ($\mu_{JT} > 0$); outside that region, the expansion produces heating. The gas–liquid coexistence curve is shown by the blue line, terminating at the critical point (the solid blue circle). The dashed lines demarcates the regions where N_2 is neither a supercritical fluid, a liquid, nor a gas.

Physical Mechanism

There are two factors that can change the temperature of a fluid during an adiabatic expansion: a change in internal energy or the conversion between potential and kinetic internal energy. Temperature is the measure of thermal kinetic energy (energy associated with molecular motion); so a change in temperature indicates a change in thermal kinetic energy. The internal energy is the sum of thermal kinetic energy and thermal potential energy. Thus, even if the internal energy does not change, the temperature can change due to conversion between kinetic and potential energy; this is what happens in a free expansion and typically produces a decrease in temperature as the fluid expands. If work is done on or by the fluid as it expands, then the total internal energy changes. This is what happens in a Joule–Thomson expansion and can produce larger heating or cooling than observed in a free expansion.

In a Joule–Thomson expansion the enthalpy remains constant. The enthalpy, H, is defined as:

$$H = U + PV$$

Where U is internal energy, V is pressure, and P is volume. Under the conditions of a Joule–Thomson expansion, the change in PV represents the work done by the fluid. If PV increases, with H constant, then U must decrease as a result of the fluid doing work on its surroundings. This produces a decrease in temperature and results in a positive Joule–Thomson coefficient. Conversely, a decrease in PV means that work is done on the fluid and the internal energy increases. If the increase in kinetic energy exceeds the increase in potential energy, there will be an increase in the temperature of the fluid and the Joule–Thomson coefficient will be negative.

For an ideal gas, PV does not change during a Joule–Thomson expansion. As a result, there is no change in internal energy; since there is also no change in thermal potential energy, there can be no change in thermal kinetic energy and, therefore, no change in temperature. In real gases, PV does change.

The ratio of the value of PV to that expected for an ideal gas at the same temperature is called the compressibility factor, Z. For a gas, this is typically less than unity at low temperature and greater than unity at high temperature. At low pressure, the value of Z always moves towards unity as a gas expands. Thus at low temperature, Z and PV will increase as the gas expands, resulting in a positive Joule–Thomson coefficient. At high temperature, Z and PV decrease as the gas expands; if the decrease is large enough, the Joule–Thomson coefficient will be negative.

For liquids, and for supercritical fluids under high pressure, PV increases as pressure increases. This is due to molecules being forced together, so that the volume can barely decrease due to higher pressure. Under such conditions, the Joule–Thomson coefficient is negative.

The physical mechanism associated with the Joule–Thomson effect is closely related to that of a shock wave, although a shock wave differs in that the change in bulk kinetic energy of the gas flow is not negligible.

The Joule–Thomson (Kelvin) Coefficient

The rate of change of temperature T with respect to pressure P in a Joule–Thomson process (that is, at constant enthalpy H) is the Joule–Thomson (Kelvin) coefficient μ_{JT}. This coefficient can be expressed in terms of the gas's volume V, its heat capacity at constant pressure C_p, and its coefficient of thermal expansion α as:

$$\mu_{JT} = \left(\frac{\partial T}{\partial P}\right)_H = \frac{V}{C_p}(\alpha T - 1)$$

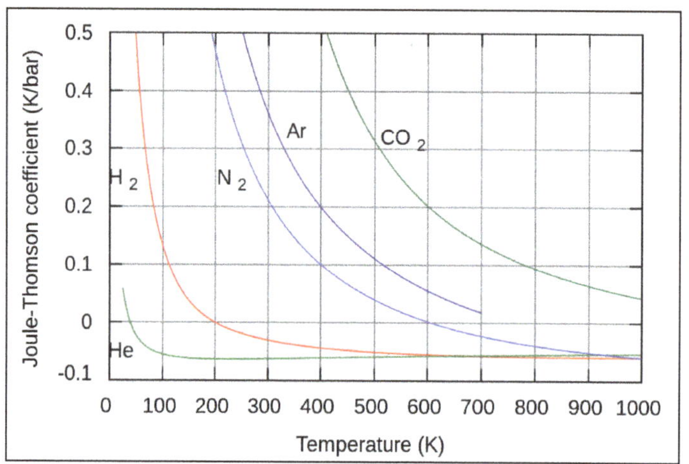

Joule–Thomson coefficients for various gases at atmospheric pressure.

The value of μ_{JT} is typically expressed in °C/bar (SI units: K/Pa) and depends on the type of gas and on the temperature and pressure of the gas before expansion. Its pressure dependence is usually only a few percent for pressures up to 100 bar.

All real gases have an inversion point at which the value of μ_{JT} changes sign. The temperature of this point, the Joule–Thomson inversion temperature, depends on the pressure of the gas before expansion.

In a gas expansion the pressure decreases, so the sign of ∂P is negative by definition. With that in mind, the following table explains when the Joule–Thomson effect cools or warms a real gas:

If the gas temperature is	then μ_{JT} is	since ∂P is	thus ∂T must be	so the gas
below the inversion temperature	positive	always negative	negative	cools
above the inversion temperature	negative	always negative	positive	warms

Helium and hydrogen are two gases whose Joule–Thomson inversion temperatures at a pressure of one atmosphere are very low (e.g., about 45 K (−228 °C) for helium). Thus, helium and hydrogen warm when expanded at constant enthalpy at typical room temperatures. On the other hand, nitrogen and oxygen, the two most abundant gases in air, have inversion temperatures of 621 K (348 °C) and 764 K (491 °C) respectively: these gases can be cooled from room temperature by the Joule–Thomson effect.

For an ideal gas, μ_{JT} is always equal to zero: ideal gases neither warm nor cool upon being expanded at constant enthalpy.

Proof that the Specific Enthalpy Remains Constant

In thermodynamics so-called "specific" quantities are quantities per unit mass (kg) and are denoted by lower-case characters. So *h*, *u*, and *v* are the specific enthalpy, specific internal energy, and specific volume (volume per unit mass, or reciprocal density), respectively. In a Joule–Thomson process the specific enthalpy *h* remains constant. To prove this, the first step is to compute the net work done when a mass *m* of the gas moves through the plug. This amount of gas has a volume of $V_1 = m\, v_1$ in the region at pressure P_1 (region 1) and a volume $V_2 = m\, v_2$ when in the region at pressure P_2 (region 2). Then in region 1, the "flow work" done *on* the amount of gas by the rest of the gas is: $W_1 = m\, P_1 v_1$. In region 2, the work done *by* the amount of gas on the rest of the gas is: $W_2 = m\, P_2 v_2$. So, the total work done *on* the mass *m* of gas is:

$$W = mP_1v_1 - mP_2v_2.$$

The change in internal energy minus the total work done on the amount of gas is, by the first law of thermodynamics, the total heat supplied to the amount of gas:

$$U - W = Q$$

In the Joule–Thomson process, the gas is insulated, so no heat is absorbed. This means that,

$$(mu_2 - mu_1) - (mP_1v_1 - mP_2v_2) = 0$$
$$mu_1 + mP_1v_1 = mu_2 + mP_2v_2$$
$$u_1 + P_1v_1 = u_2 + P_2v_2$$

where u_1 and u_2 denote the specific internal energies of the gas in regions 1 and 2, respectively. Using the definition of the specific enthalpy $h = u + Pv$, the above equation implies that,

$$h_1 = h_2$$

where h_1 and h_2 denote the specific enthalpies of the amount of gas in regions 1 and 2, respectively.

Throttling in the T-s Diagram

The red dome represents the two-phase region with the low-entropy side (the saturated liquid) and the high-entropy side (the saturated gas). The black curves give the *T*-*s* relation along isobars. The pressures are indicated in bar. The blue curves are isenthalps (curves of constant specific enthalpy). The specific enthalpies are indicated in kJ/kg. The specific points a, b, etc., are treated in the main text.

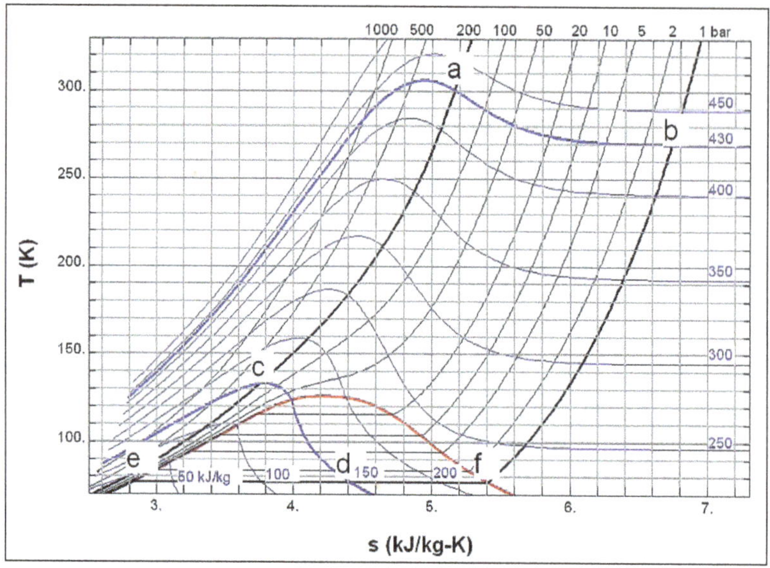

T-s diagram of nitrogen.

A very convenient way to get a quantitative understanding of the throttling process is by using diagrams. There are many types of diagrams (*h-T* diagram, *h-P* diagram, etc.) Commonly used are the so-called *T-s* diagrams. Figure shows the *T-s* diagram of nitrogen as an example. Various points are indicated as follows:

- $T = 300$ K, $p = 200$ bar, $s = 5.16$ kJ/(kgK), $h = 430$ kJ/kg;
- $T = 270$ K, $p = 1$ bar, $s = 6.79$ kJ/(kgK), $h = 430$ kJ/kg;
- $T = 133$ K, $p = 200$ bar, $s = 3.75$ kJ/(kgK), $h = 150$ kJ/kg;
- $T = 77.2$ K, $p = 1$ bar, $s = 4.40$ kJ/(kgK), $h = 150$ kJ/kg;
- $T = 77.2$ K, $p = 1$ bar, $s = 2.83$ kJ/(kgK), $h = 28$ kJ/kg (saturated liquid at 1 bar);
- $T = 77.2$ K, $p = 1$ bar, $s = 5.41$ kJ/(kgK), $h =230$ kJ/kg (saturated gas at 1 bar).

As shown before, throttling keeps h constant. E.g. throttling from 200 bar and 300 K follows the isenthalp (line of constant specific enthalpy) of 430 kJ/kg. At 1 bar it results in point b which has a temperature of 270 K. So throttling from 200 bar to 1 bar gives a cooling from room temperature to below the freezing point of water. Throttling from 200 bar and an initial temperature of 133 K to 1 bar results in point d, which is in the two-phase region of nitrogen at a temperature of 77.2 K. Since the enthalpy is an extensive parameter the enthalpy in d (h_d) is equal to the enthalpy in e (h_e) multiplied with the mass fraction of the liquid in d (x_d) plus the enthalpy in f (h_f) multiplied with the mass fraction of the gas in d ($1 - x_d$). So

$$h_d = x_d h_e + (1 - x_d) h_f.$$

With numbers: $150 = x_d\, 28 + (1 - x_d)\, 230$ so x_d is about 0.40. This means that the mass fraction of the liquid in the liquid–gas mixture leaving the throttling valve is 40%.

Derivation of the Joule–Thomson Coefficient

It is difficult to think physically about what the Joule–Thomson coefficient, μ_{JT}, represents. Also, modern determinations of μ_{JT} do not use the original method used by Joule and Thomson, but instead measure a different, closely related quantity. Thus, it is useful to derive relationships between μ_{JT} and other, more convenient quantities.

The first step in obtaining these results is to note that the Joule–Thomson coefficient involves the three variables *T*, *P*, and *H*. A useful result is immediately obtained by applying the cyclic rule; in terms of these three variables that rule may be written:

$$\left(\frac{\partial T}{\partial P}\right)_H \left(\frac{\partial H}{\partial T}\right)_P \left(\frac{\partial P}{\partial H}\right)_T = -1.$$

Each of the three partial derivatives in this expression has a specific meaning. The first is μ_{JT}, the second is the constant pressure heat capacity, C_{p}, defined by,

$$C_{\mathrm{p}} = \left(\frac{\partial H}{\partial T}\right)_P$$

and the third is the inverse of the *isothermal Joule–Thomson coefficient*, μ_{T}, defined by,

$$\mu_{\mathrm{T}} = \left(\frac{\partial H}{\partial P}\right)_T.$$

This last quantity is more easily measured than μ_{JT} . Thus, the expression from the cyclic rule becomes,

$$\mu_{\mathrm{JT}} = -\frac{\mu_{\mathrm{T}}}{C_p}.$$

This equation can be used to obtain Joule–Thomson coefficients from the more easily measured isothermal Joule–Thomson coefficient. It is used in the following to obtain a mathematical expression for the Joule–Thomson coefficient in terms of the volumetric properties of a fluid.

To proceed further, the starting point is the fundamental equation of thermodynamics in terms of enthalpy; this is,

$$\mathrm{d}H = T\mathrm{d}S + V\mathrm{d}P.$$

Now "dividing through" by dP, while holding temperature constant, yields,

$$\left(\frac{\partial H}{\partial P}\right)_T = T\left(\frac{\partial S}{\partial P}\right)_T + V$$

The partial derivative on the left is the isothermal Joule–Thomson coefficient, μ_{T}, and the one on the right can be expressed in terms of the coefficient of thermal expansion via a Maxwell relation.

The appropriate relation is,

$$\left(\frac{\partial S}{\partial P}\right)_T = -\left(\frac{\partial V}{\partial T}\right)_P = -V\alpha$$

where α is the cubic coefficient of thermal expansion. Replacing these two partial derivatives yields,

$$\mu_{\mathrm{T}} = -TV\alpha + V.$$

This expression can now replace μ_{T} in the earlier equation for μ_{JT} to obtain:

$$\mu_{\mathrm{JT}} \equiv \left(\frac{\partial T}{\partial P}\right)_H = \frac{V}{C_{\mathrm{p}}}(\alpha T - 1)$$

This provides an expression for the Joule–Thomson coefficient in terms of the commonly available properties heat capacity, molar volume, and thermal expansion coefficient. It shows that the Joule–Thomson inversion temperature, at which μ_{JT} is zero, occurs when the coefficient of thermal expansion is equal to the inverse of the temperature. Since this is true at all temperatures for ideal gases, the Joule–Thomson coefficient of an ideal gas is zero at all temperatures.

Joule's Second Law

It is easy to verify that for an ideal gas defined by suitable microscopic postulates that $\alpha T = 1$, so the temperature change of such an ideal gas at a Joule–Thomson expansion is zero. For such an ideal gas, this theoretical result implies that:

The internal energy of a fixed mass of an ideal gas depends only on its temperature (not pressure or volume).

This rule was originally found by Joule experimentally for real gases and is known as Joule's second law. More refined experiments of course found important deviations from it.

Temperature Inversion

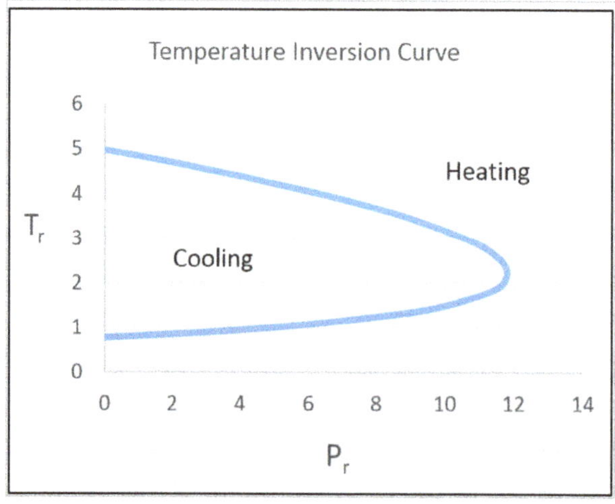

The temperature of a gas at which the reduction in pressure does not lead to any change in temperature is known as inversion temperature. The gas gets heated up on expansion and cools down below this temperature.

Applications of Joule-Thomson Effect

- The cooling produced in the Joule-Thomson expansion has made it a very valuable tool in refrigeration.
- The effect is applied in the Linde technique in the petrochemical industry, where the cooling effect is used to liquefy gases.
- It is also used in many cryogenic applications. For example for the production of liquid nitrogen, oxygen, and argon.
- The effect can also be used to liquefy even helium.

References

- Mayers-relation-mayers-formula, ideal-gas-law: nuclear-power.net, Retrieved 17 August, 2019
- Meyers-relation, kinetic-theory-of-gases: askiitians.com, Retrieved 25 January, 2019
- Maxwells-relations: byjus.com, Retrieved 02 April, 2019
- Bridgman-formulas: eoht.info, Retrieved 04 May, 2019
- Clausius-clapeyron-equation: glossary.ametsoc.org, Retrieved 25 June, 2019
- Antoine-equation: m.tau.ac.il, Retrieved 14 February, 2019
- Derive-the-gibbs-duhem-equation-and-give-its-applications, second-law-of-thermodynamics: thebigger.com, Retrieved 11 August, 2019
- Joule-thomson-effect: byjus.com, Retrieved 29 June, 2019

Permissions

All chapters in this book are published with permission under the Creative Commons Attribution Share Alike License or equivalent. Every chapter published in this book has been scrutinized by our experts. Their significance has been extensively debated. The topics covered herein carry significant information for a comprehensive understanding. They may even be implemented as practical applications or may be referred to as a beginning point for further studies.

We would like to thank the editorial team for lending their expertise to make the book truly unique. They have played a crucial role in the development of this book. Without their invaluable contributions this book wouldn't have been possible. They have made vital efforts to compile up to date information on the varied aspects of this subject to make this book a valuable addition to the collection of many professionals and students.

This book was conceptualized with the vision of imparting up-to-date and integrated information in this field. To ensure the same, a matchless editorial board was set up. Every individual on the board went through rigorous rounds of assessment to prove their worth. After which they invested a large part of their time researching and compiling the most relevant data for our readers.

The editorial board has been involved in producing this book since its inception. They have spent rigorous hours researching and exploring the diverse topics which have resulted in the successful publishing of this book. They have passed on their knowledge of decades through this book. To expedite this challenging task, the publisher supported the team at every step. A small team of assistant editors was also appointed to further simplify the editing procedure and attain best results for the readers.

Apart from the editorial board, the designing team has also invested a significant amount of their time in understanding the subject and creating the most relevant covers. They scrutinized every image to scout for the most suitable representation of the subject and create an appropriate cover for the book.

The publishing team has been an ardent support to the editorial, designing and production team. Their endless efforts to recruit the best for this project, has resulted in the accomplishment of this book. They are a veteran in the field of academics and their pool of knowledge is as vast as their experience in printing. Their expertise and guidance has proved useful at every step. Their uncompromising quality standards have made this book an exceptional effort. Their encouragement from time to time has been an inspiration for everyone.

The publisher and the editorial board hope that this book will prove to be a valuable piece of knowledge for students, practitioners and scholars across the globe.

Index

www.ingramcontent.com/pod-product-compliance
Lightning Source LLC
Chambersburg PA
CBHW082036190326
41458CB00010B/3381

* 9 7 8 1 6 4 1 7 2 6 5 9 7 *